SIDONIA 4

U NIHEI

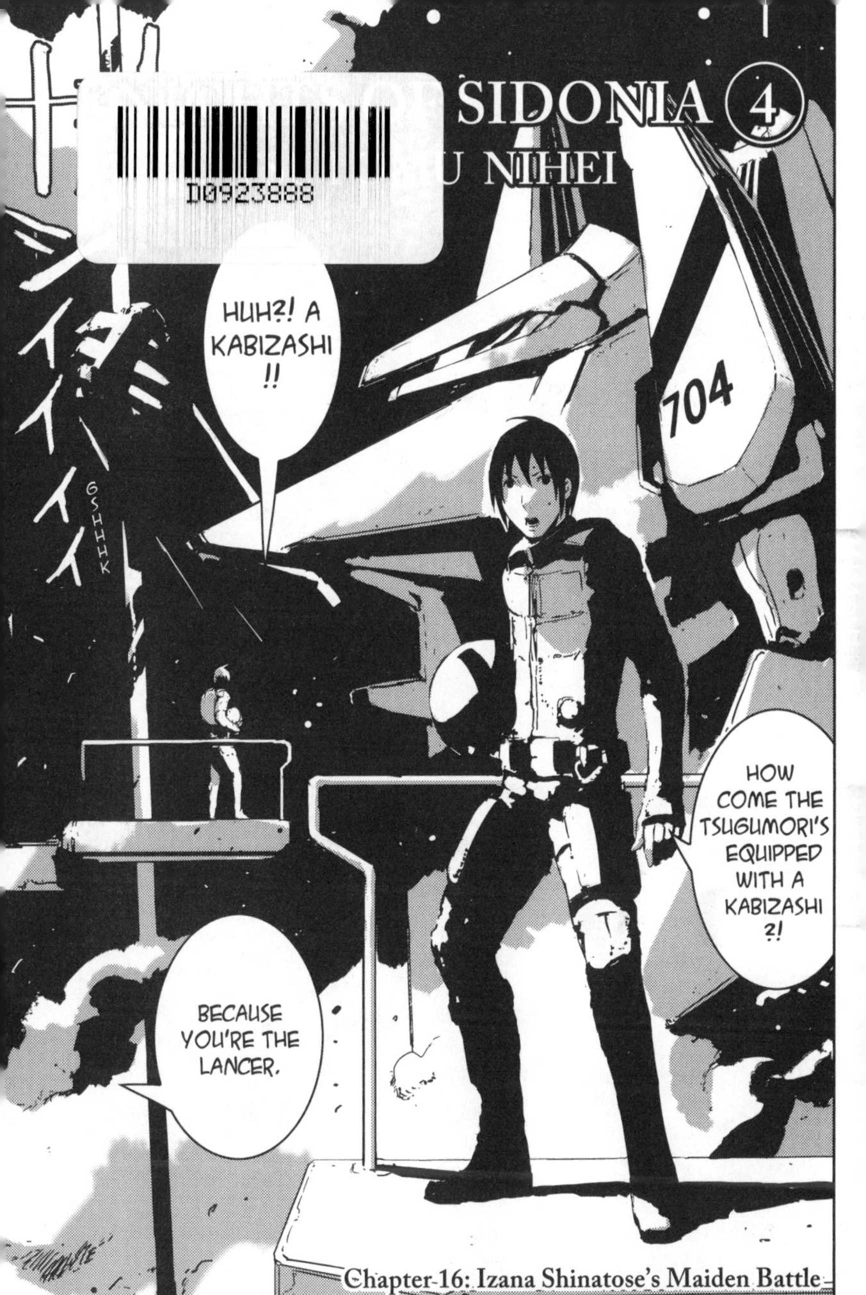
HUH?! A KABIZASHI !!
GSHHHK
704
HOW COME THE TSUGUMORI'S EQUIPPED WITH A KABIZASHI ?!
BECAUSE YOU'RE THE LANCER.

Chapter-16: Izana Shinatose's Maiden Battle

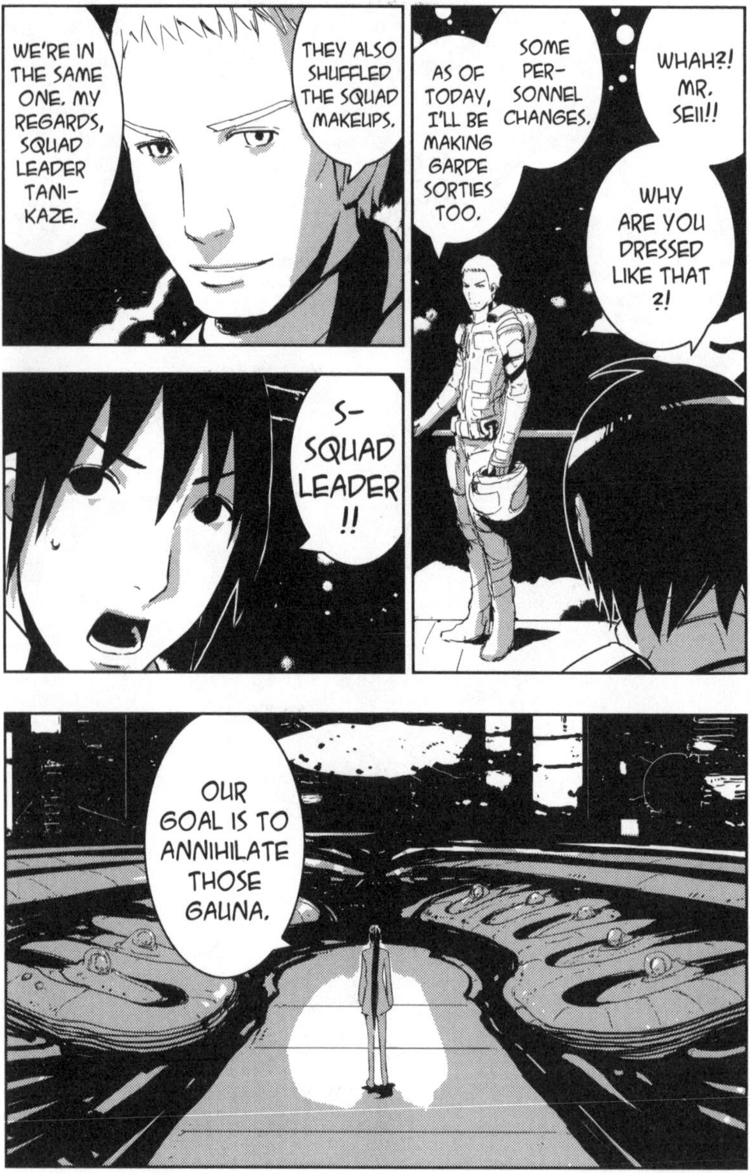
WHAH?! MR. SEII!!
WHY ARE YOU DRESSED LIKE THAT?!
SOME PERSONNEL CHANGES.
AS OF TODAY, I'LL BE MAKING GARDE SORTIES TOO.
THEY ALSO SHUFFLED THE SQUAD MAKEUPS.
WE'RE IN THE SAME ONE. MY REGARDS, SQUAD LEADER TANIKAZE.
S-SQUAD LEADER!!
OUR GOAL IS TO ANNIHILATE THOSE GAUNA.

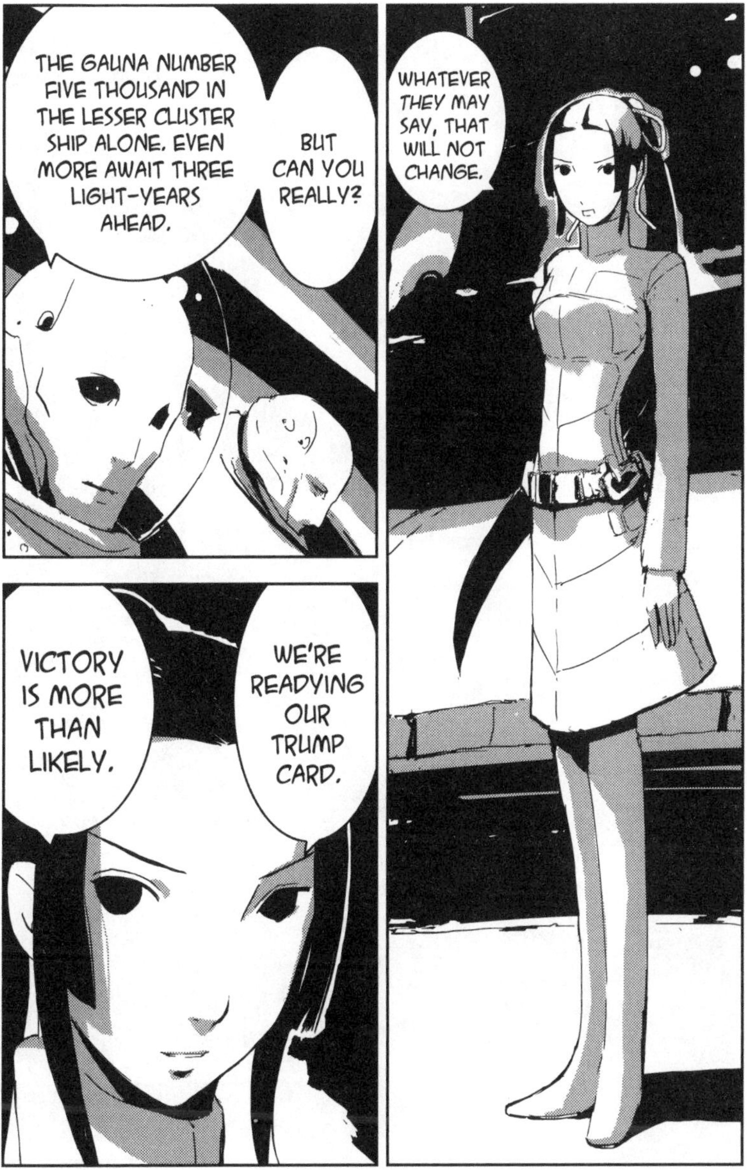
WHATEVER THEY MAY SAY, THAT WILL NOT CHANGE.
BUT CAN YOU REALLY?
THE GAUNA NUMBER FIVE THOUSAND IN THE LESSER CLUSTER SHIP ALONE. EVEN MORE AWAIT THREE LIGHT-YEARS AHEAD.
WE'RE READYING OUR TRUMP CARD.
VICTORY IS MORE THAN LIKELY.

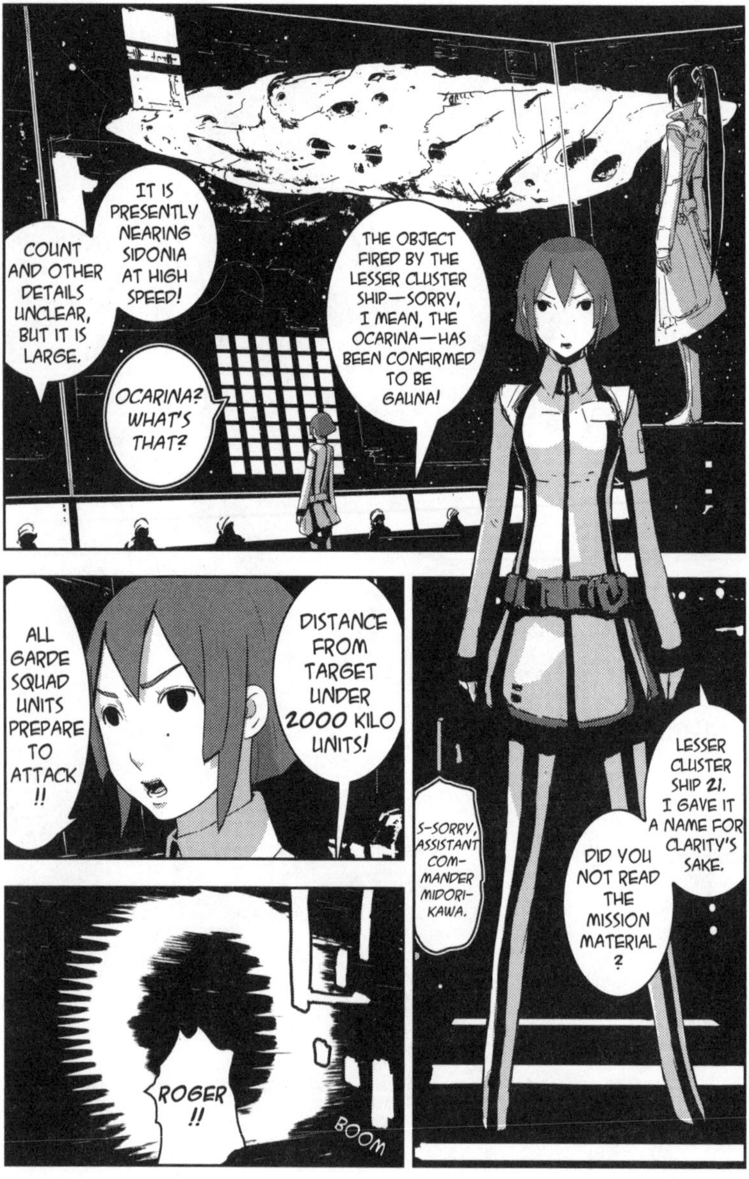
THE OBJECT FIRED BY THE LESSER CLUSTER SHIP—SORRY, I MEAN, THE OCARINA—HAS BEEN CONFIRMED TO BE GAUNA!
IT IS PRESENTLY NEARING SIDONIA AT HIGH SPEED!
COUNT AND OTHER DETAILS UNCLEAR, BUT IT IS LARGE.
OCARINA? WHAT'S THAT?
LESSER CLUSTER SHIP 21. I GAVE IT A NAME FOR CLARITY'S SAKE.
DID YOU NOT READ THE MISSION MATERIAL?
S-SORRY, ASSISTANT COM-MANDER MIDORI-KAWA.
DISTANCE FROM TARGET UNDER 2000 KILO UNITS!
ALL GARDE SQUAD UNITS PREPARE TO ATTACK!!
ROGER!!
BOOM

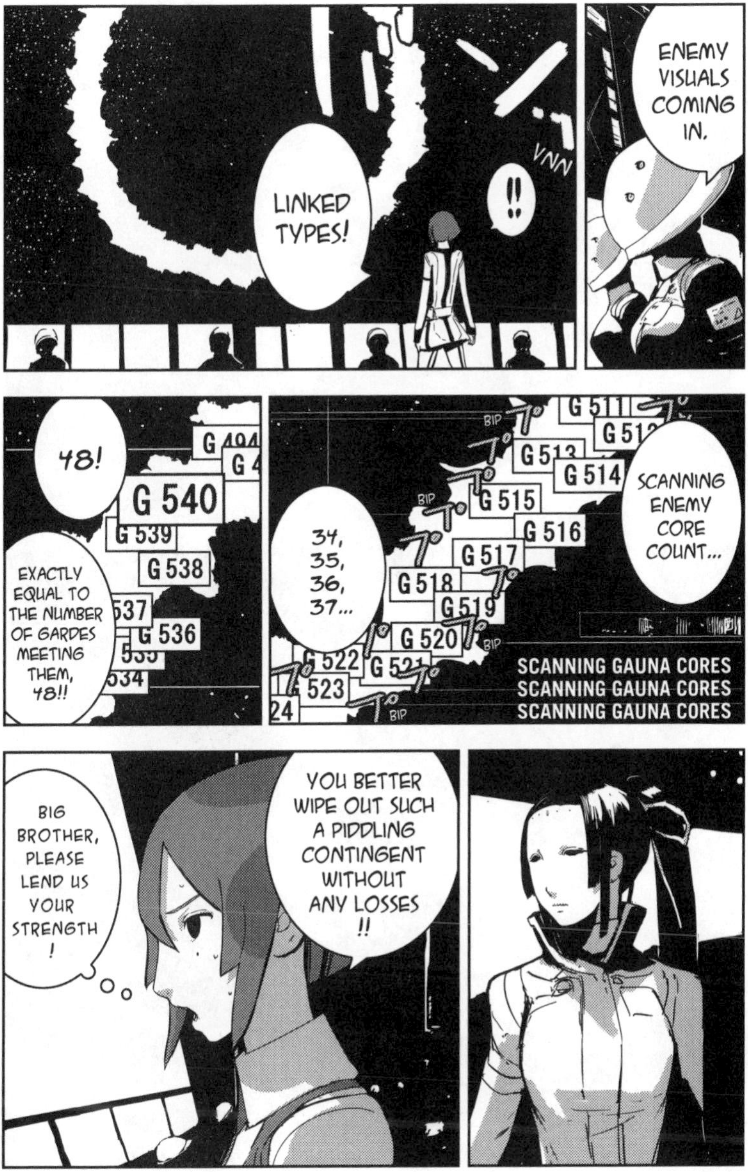
ENEMY VISUALS COMING IN.
VNN
!!
LINKED TYPES!
SCANNING ENEMY CORE COUNT...
G 514
G 515
G 516
G 517
G 518
G 519
G 520
G 523
BIP
34, 35, 36, 37...
SCANNING GAUNA CORES
SCANNING GAUNA CORES
SCANNING GAUNA CORES
48!
G 540
G 539
G 538
G 536
EXACTLY EQUAL TO THE NUMBER OF GARDES MEETING THEM, 48!!
YOU BETTER WIPE OUT SUCH A PIDDLING CONTINGENT WITHOUT ANY LOSSES!!
BIG BROTHER, PLEASE LEND US YOUR STRENGTH!

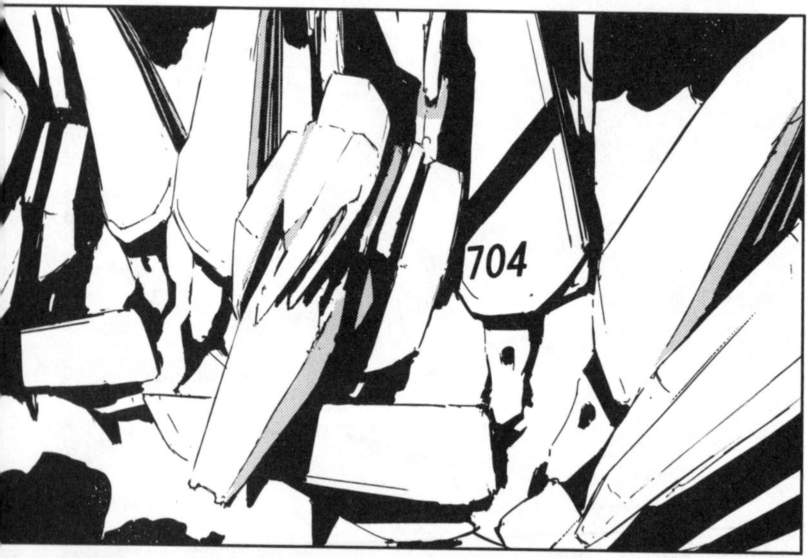
704
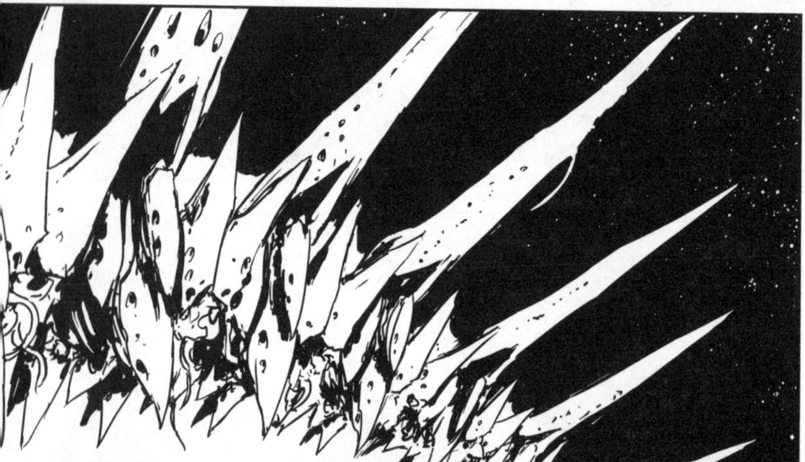

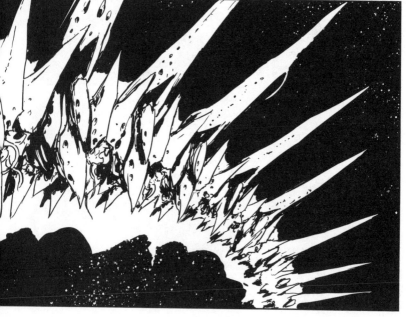

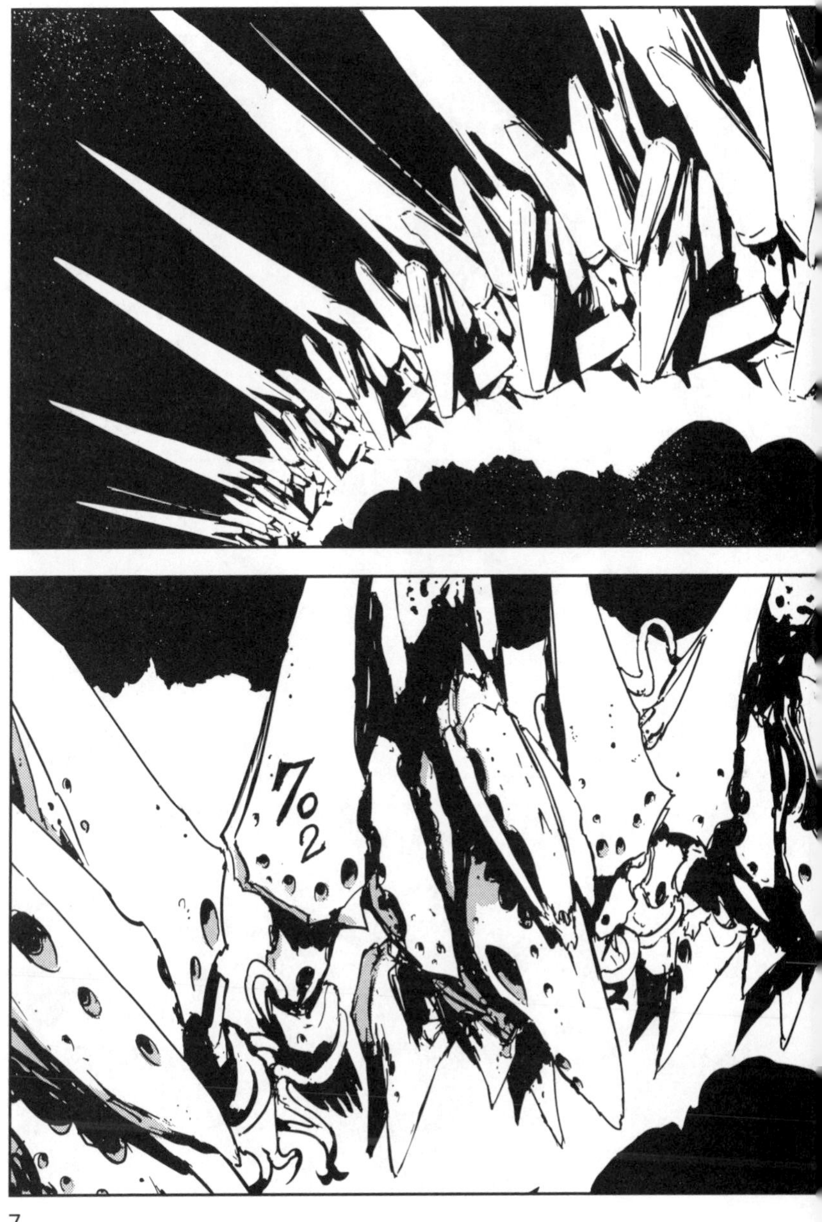
702

G 490

G490, THE HAWK MOTH, CONFIRMED AMONG ENEMY RANKS!

THIS GAUNA!!

HOSHI-JIRO...

SO IT'S HERE...

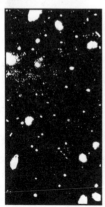

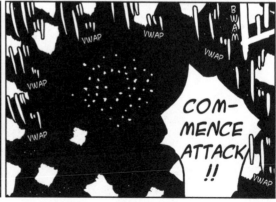

VWAP

VWAP

VWAP

VWAP

BWAM

VWAP

COM-MENCE ATTACK!!

VWAP

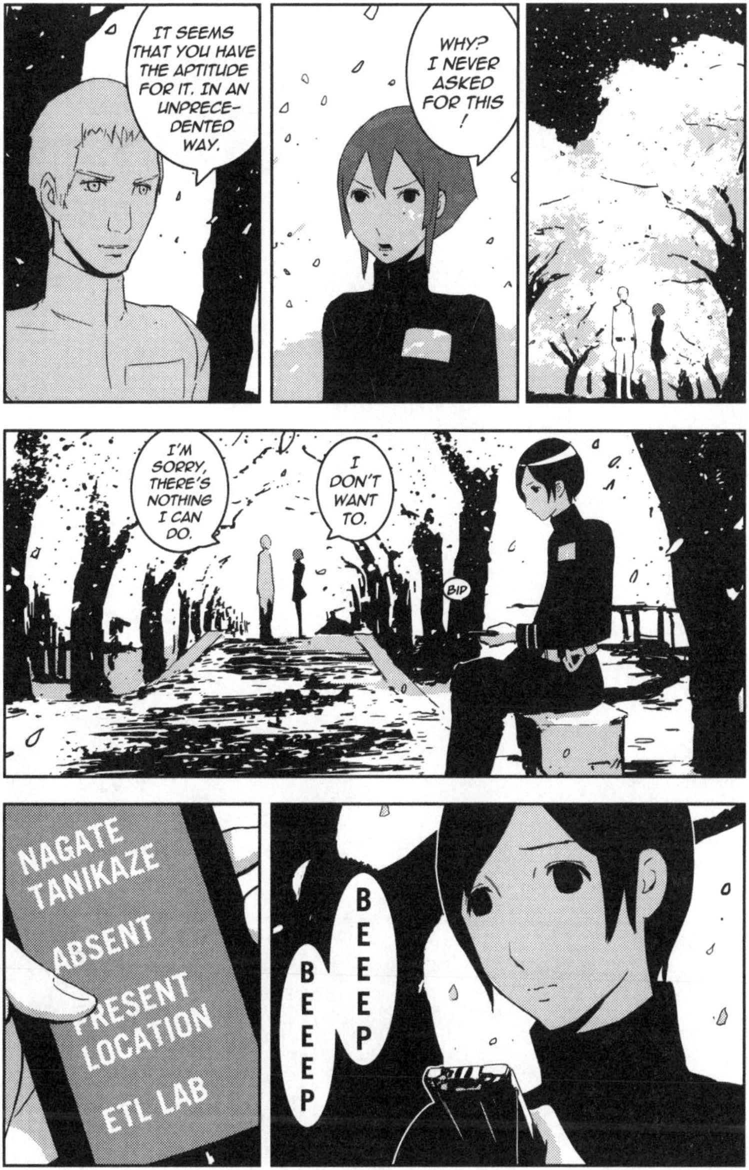
WHY? I NEVER ASKED FOR THIS!
IT SEEMS THAT YOU HAVE THE APTITUDE FOR IT. IN AN UNPRECE-DENTED WAY.
I DON'T WANT TO.
I'M SORRY, THERE'S NOTHING I CAN DO.
BIP
BEEEP
BEEEP
NAGATE TANIKAZE
ABSENT
PRESENT LOCATION
ETL LAB

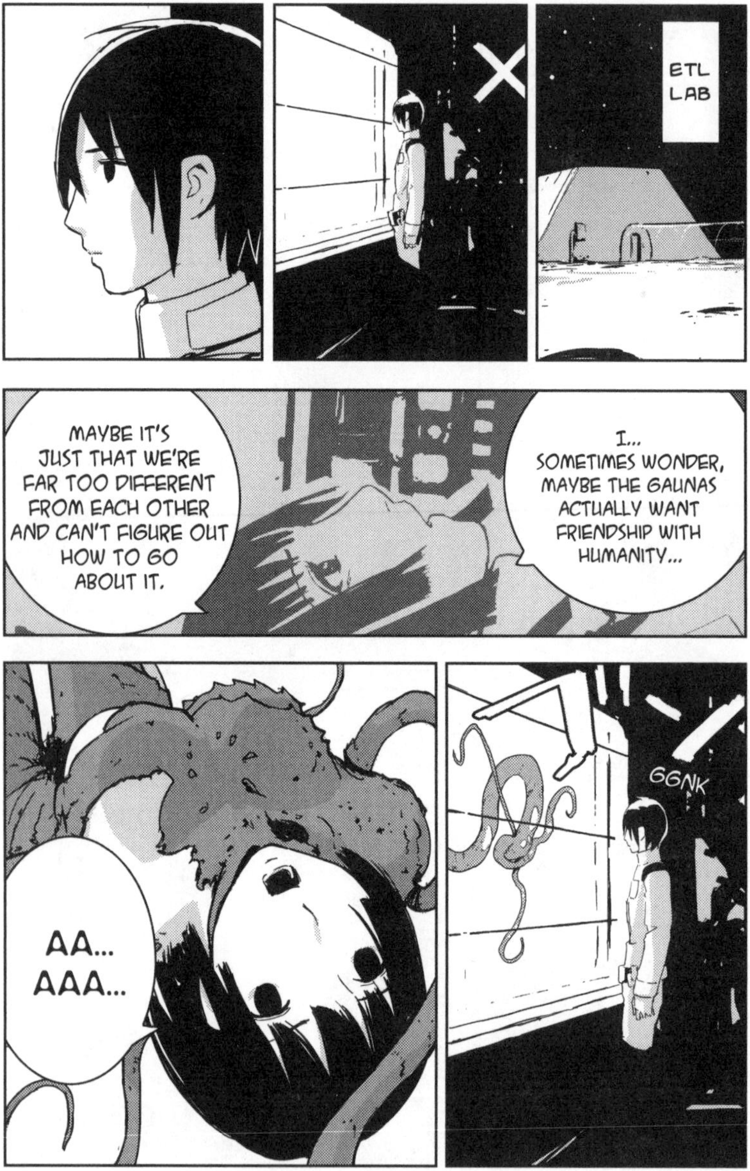
ETL LAB
I... SOMETIMES WONDER, MAYBE THE GAUNAS ACTUALLY WANT FRIENDSHIP WITH HUMANITY...
MAYBE IT'S JUST THAT WE'RE FAR TOO DIFFERENT FROM EACH OTHER AND CAN'T FIGURE OUT HOW TO GO ABOUT IT.
GGNK
AA... AAA...

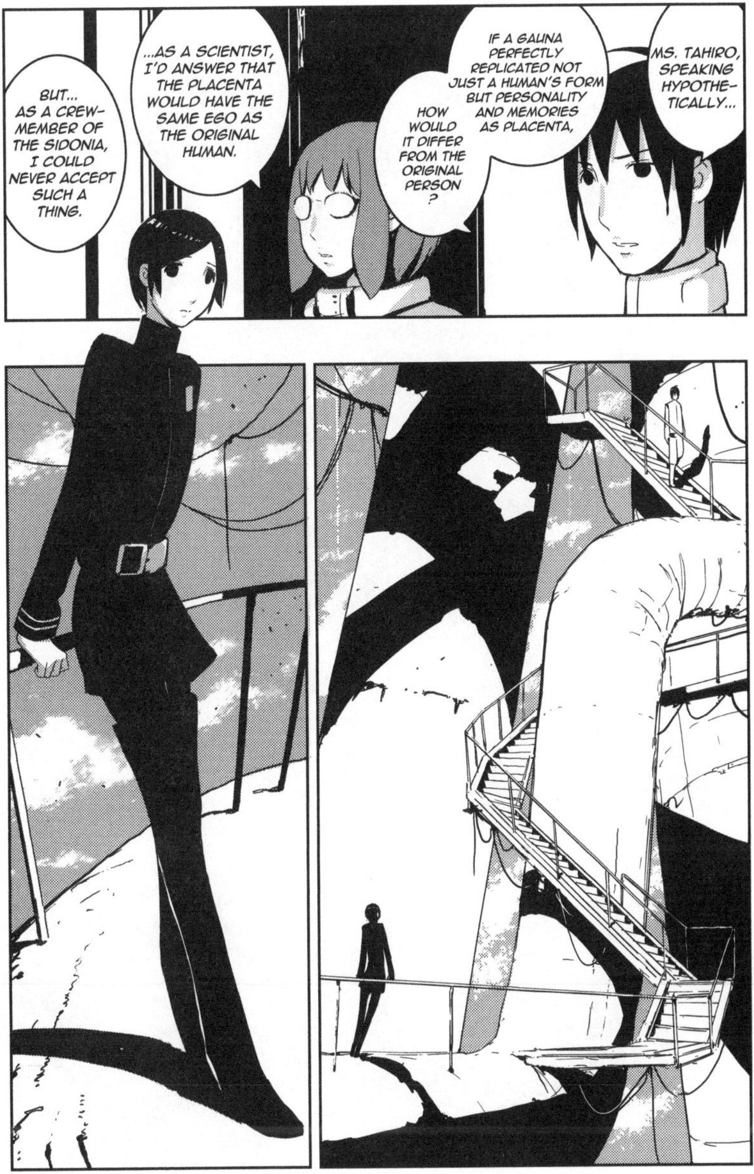
MS. TAHIRO, SPEAKING HYPOTHETICALLY...
IF A GAUNA PERFECTLY REPLICATED NOT JUST A HUMAN'S FORM BUT PERSONALITY AND MEMORIES AS PLACENTA,
HOW WOULD IT DIFFER FROM THE ORIGINAL PERSON?
...AS A SCIENTIST, I'D ANSWER THAT THE PLACENTA WOULD HAVE THE SAME EGO AS THE ORIGINAL HUMAN.
BUT... AS A CREW-MEMBER OF THE SIDONIA, I COULD NEVER ACCEPT SUCH A THING.

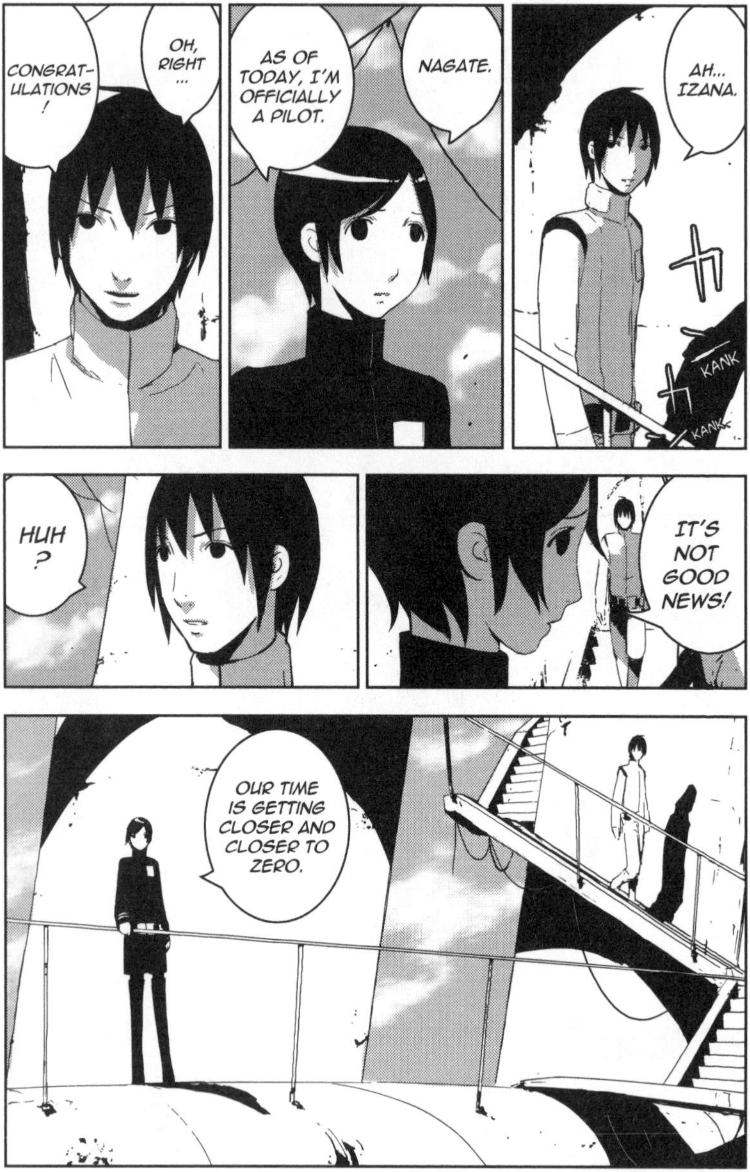
AH... IZANA.
KANK
KANK
NAGATE.
AS OF TODAY, I'M OFFICIALLY A PILOT.
OH, RIGHT ...
CONGRAT-ULATIONS!
IT'S NOT GOOD NEWS!
HUH?
OUR TIME IS GETTING CLOSER AND CLOSER TO ZERO.

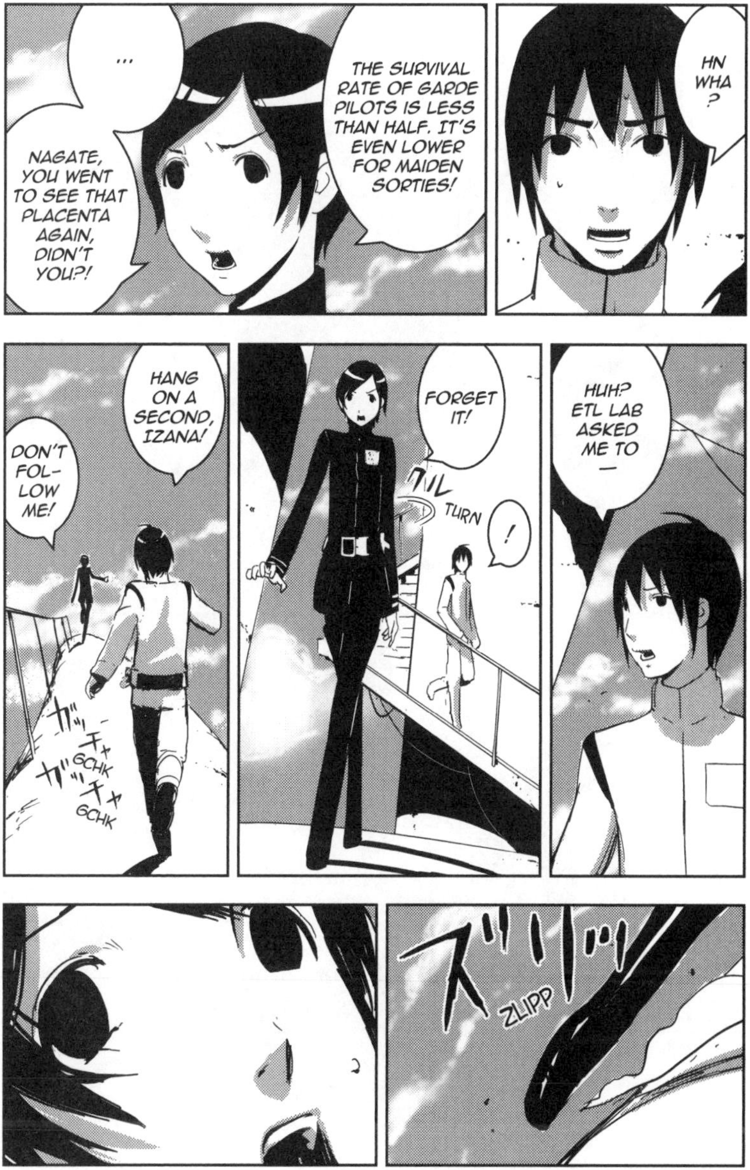
HN WHA ?
THE SURVIVAL RATE OF GARDE PILOTS IS LESS THAN HALF. IT'S EVEN LOWER FOR MAIDEN SORTIES!
...
NAGATE, YOU WENT TO SEE THAT PLACENTA AGAIN, DIDN'T YOU?!
HUH? ETL LAB ASKED ME TO —
FORGET IT!
TURN
!
HANG ON A SECOND, IZANA!
DON'T FOL-LOW ME!
GCHK
GCHK
ZLIPP

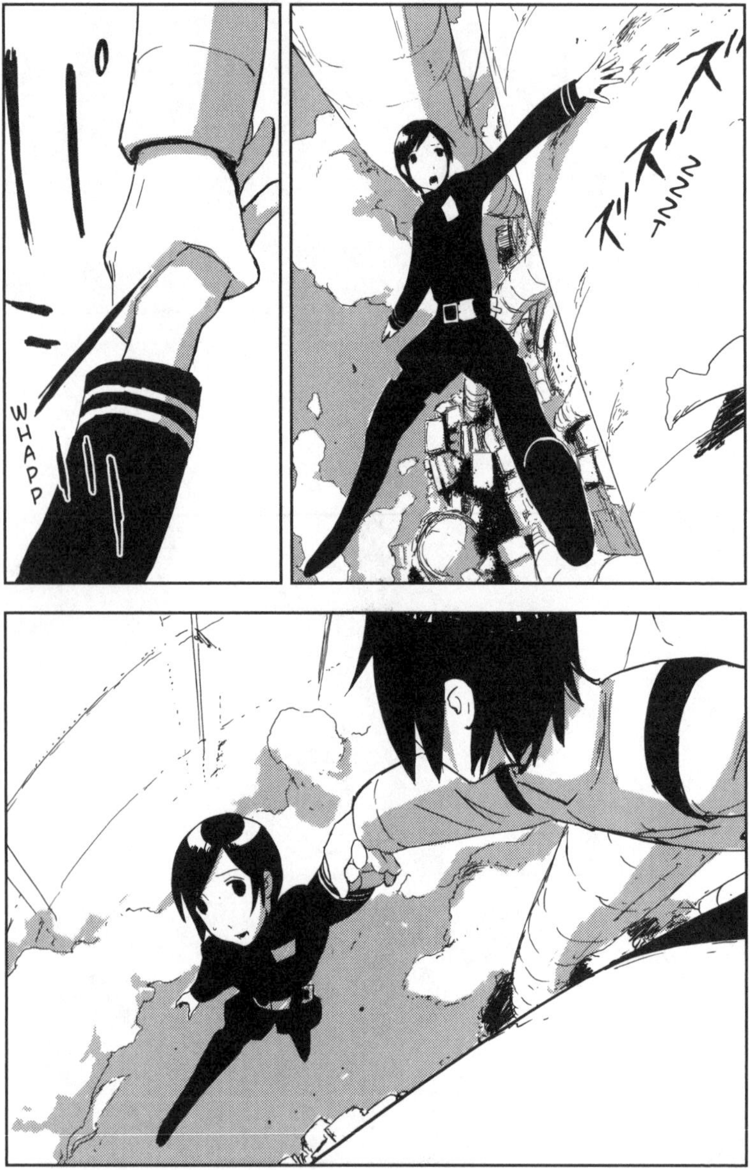
ZUNNN
WHAPP

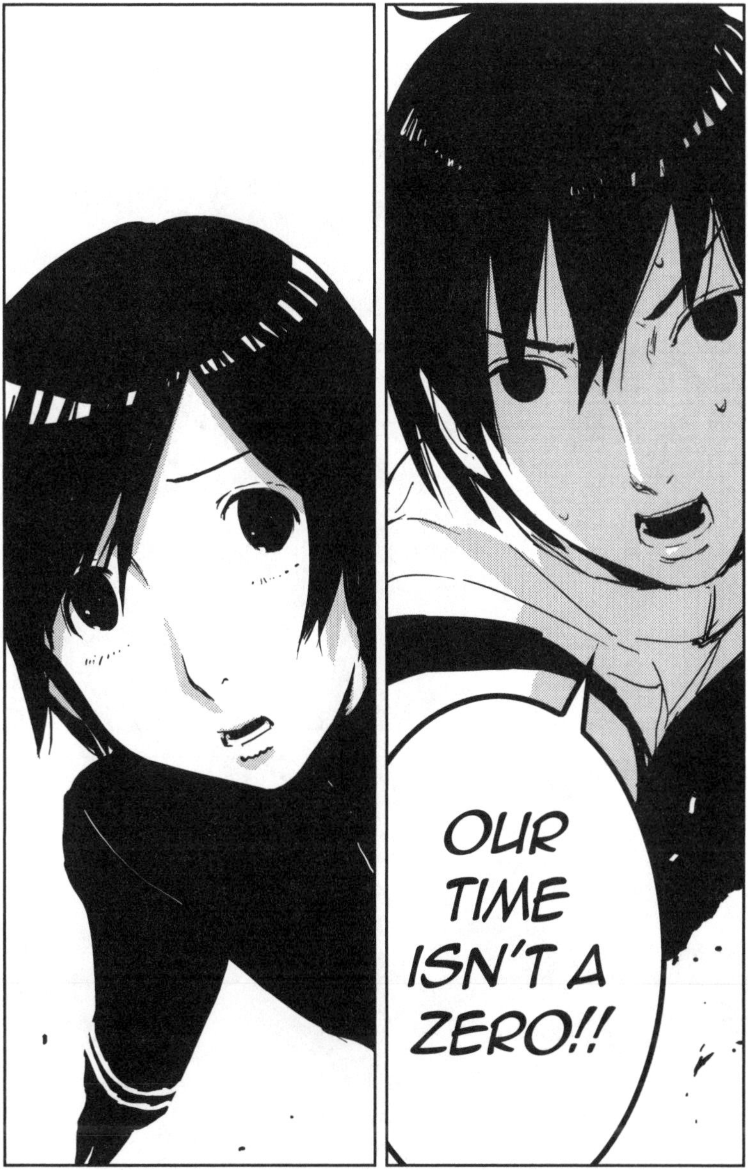
OUR TIME ISN'T A ZERO!!

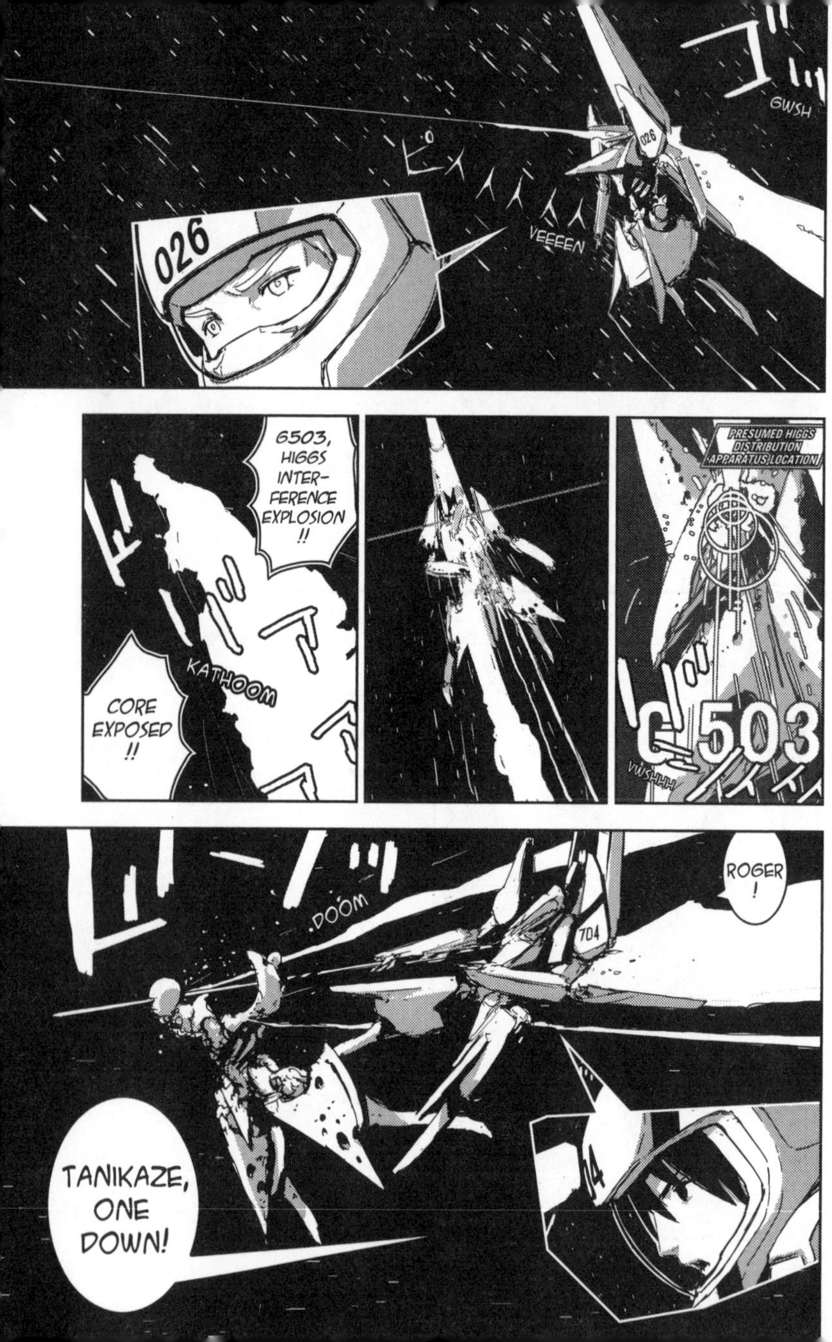
GWSH
VEEEEN
026
PRESUMED HIGGS DISTRIBUTION APPARATUS LOCATION
503
VWSHHH
6503, HIGGS INTERFERENCE EXPLOSION !!
KATHOOM
CORE EXPOSED !!
ROGER !
DOOM
704
TANIKAZE, ONE DOWN!

NUMBER 704 TANIKAZE UNIT HAS DESTROYED ITS NINE-TEENTH!

704 TANIKAZE UNIT

WHAT ARE YOU TALKING ABOUT? YOU CAME UP WITH IT.

THAT WAS AMAZING, MR. SEII!

VWOOOSH

EVEN WITH MINIMUM OUTPUT, THE INTERFERENCE EXPLOSION WILL TEAR OFF THEIR PLACENTA!!

AIM FOR THE ENEMIES' HIGGS PARTICLE CANNONS!

ALL RIGHT, WE CAN DO THIS. SEII'S TACTIC IS EFFECTIVE AGAINST GARDE-FORM GAUNAS.

SHINKI SQUAD, SUCCESSFULLY DESTROYED ONE!!

KUNATO, MOVE TO RECOVER KABIZASHI NUMBER 7!

TANIKAZE SQUAD, MOVE TO SUPPORT BETA SQUAD!

FELDT AND SHIMODA, FALL BACK!

GARDE	GAUNA
38/48	13/48

13 GAUNAS REMAINING!

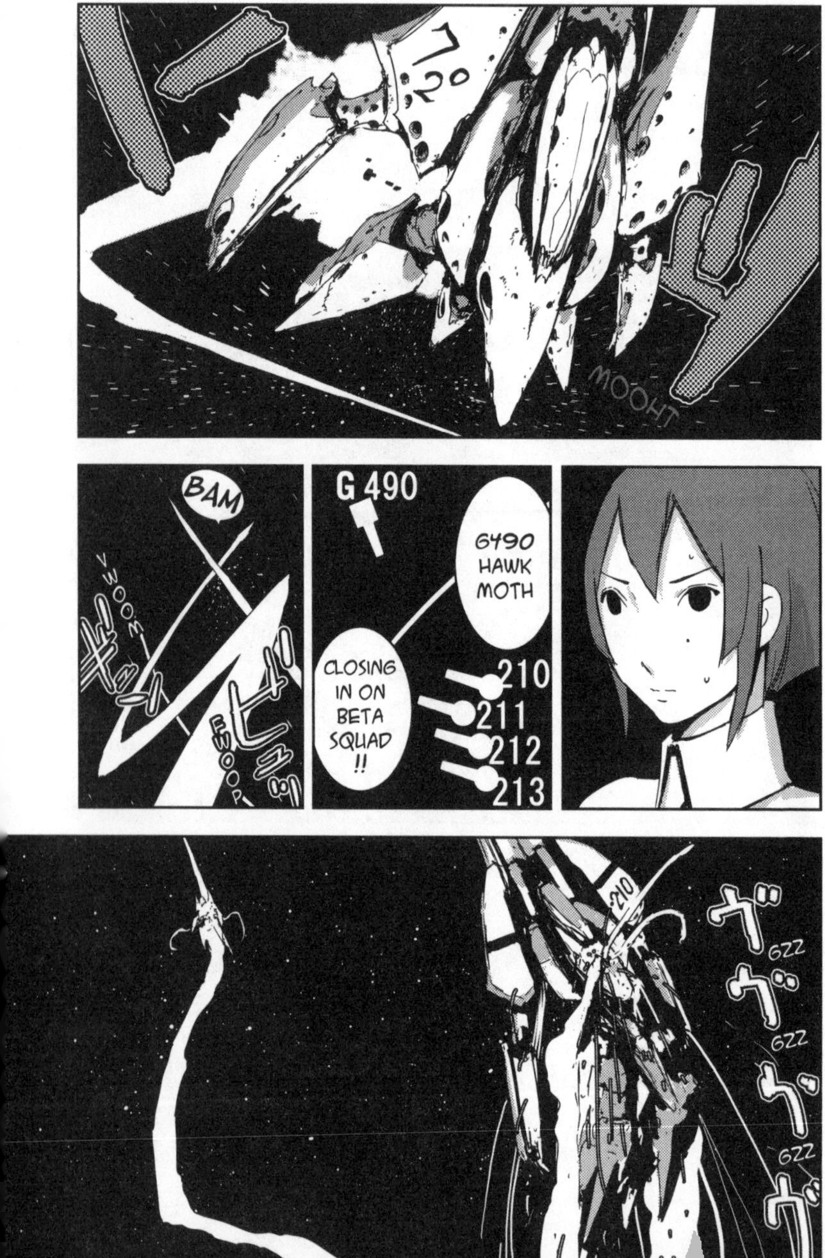
MOOHT
G 490
6490 HAWK MOTH
CLOSING IN ON BETA SQUAD !!
210
211
212
213
BAM
VWOOM
FWOOP
210
GZZ
GZZ
GZZ

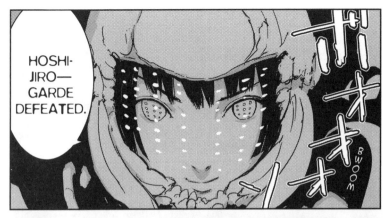

HOSHIJIRO— GARDE DEFEATED.

HOSHIJIRO!!

AAH!

UNIT 211, BEING FOLLOWED!

AAAAAAAAHH!!

HOSHIJIRO, STOP IT ALREADY!!

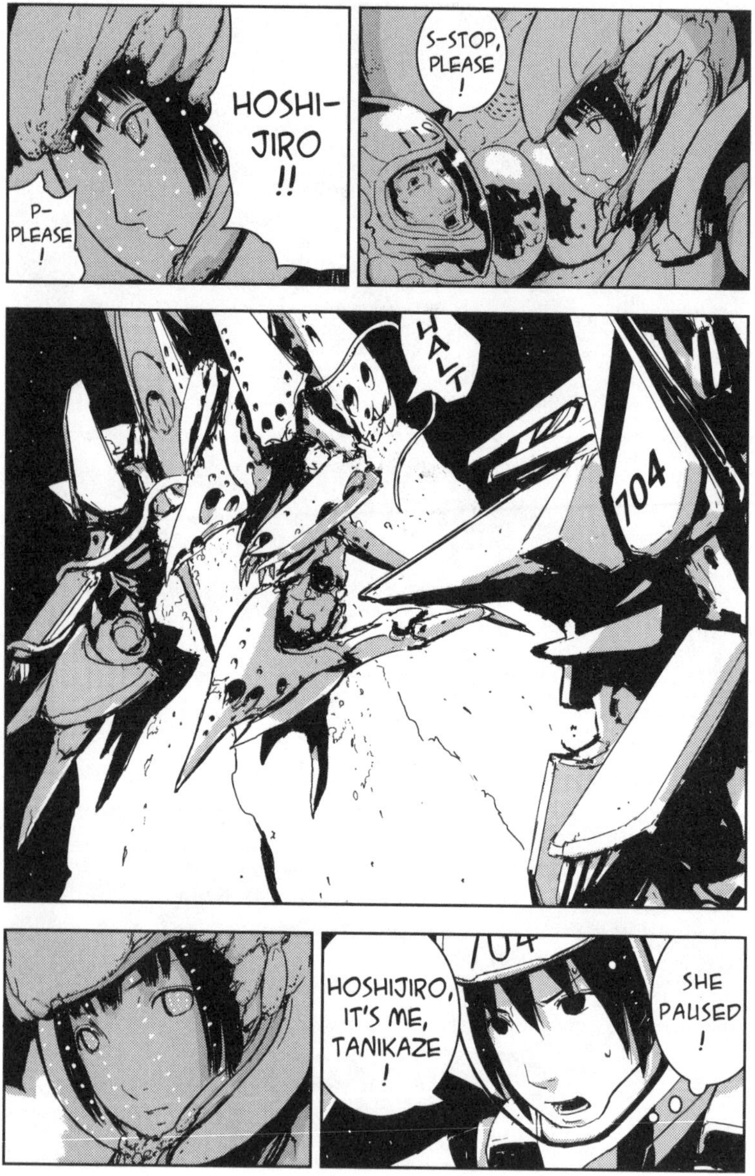
S-STOP, PLEASE!
HOSHI-JIRO!!
P-PLEASE!
HALT
SHE PAUSED!
HOSHIJIRO, IT'S ME, TANIKAZE!

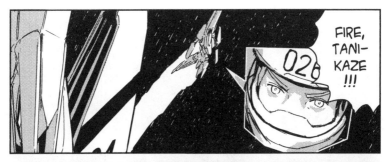

FIRE, TANI-KAZE !!!

CRLINK

PKLIK

DAMN!

!!

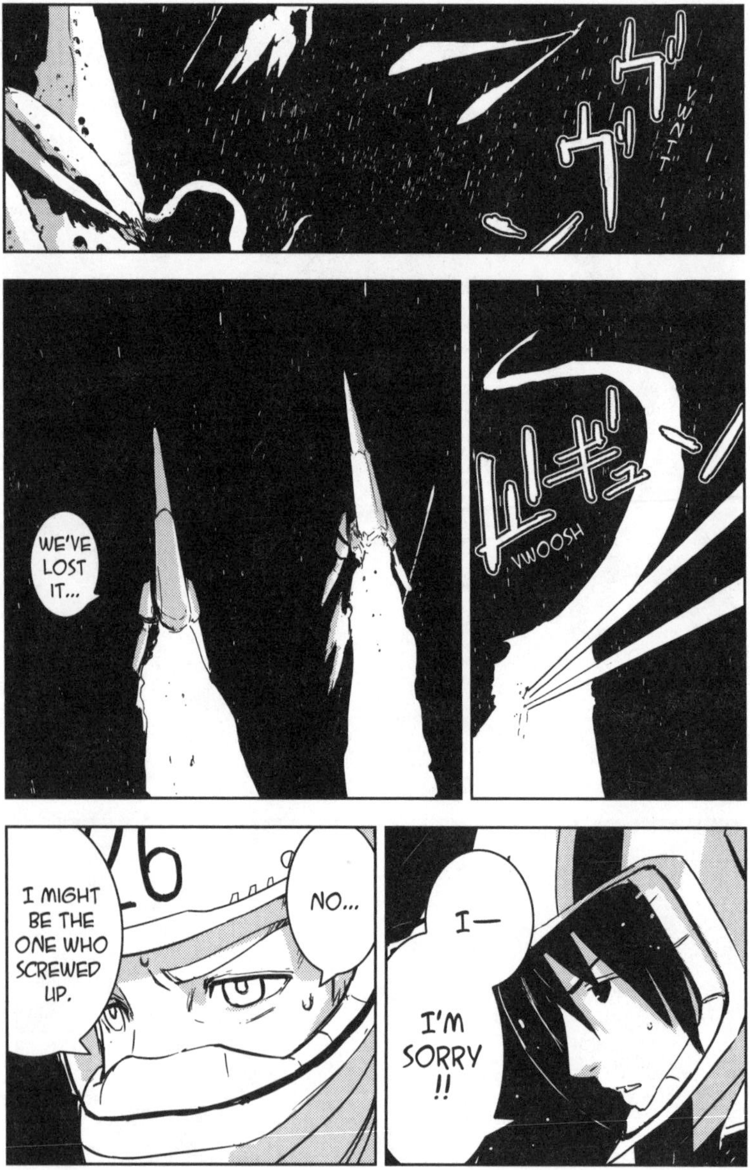
VWNTT
VWOOSH
WE'VE LOST IT...
I—
I'M SORRY !!
NO...
I MIGHT BE THE ONE WHO SCREWED UP.

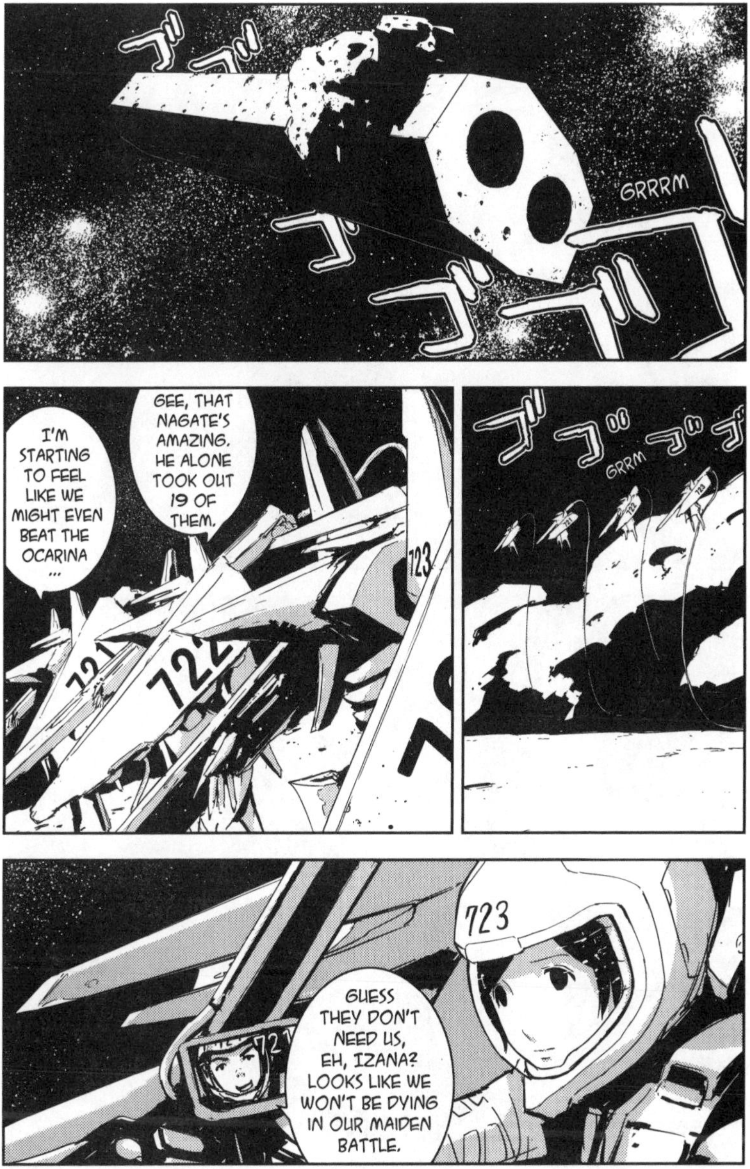
GRRRM
GRRM
GEE, THAT NAGATE'S AMAZING. HE ALONE TOOK OUT 19 OF THEM.
I'M STARTING TO FEEL LIKE WE MIGHT EVEN BEAT THE OCARINA ...
GUESS THEY DON'T NEED US, EH, IZANA? LOOKS LIKE WE WON'T BE DYING IN OUR MAIDEN BATTLE.

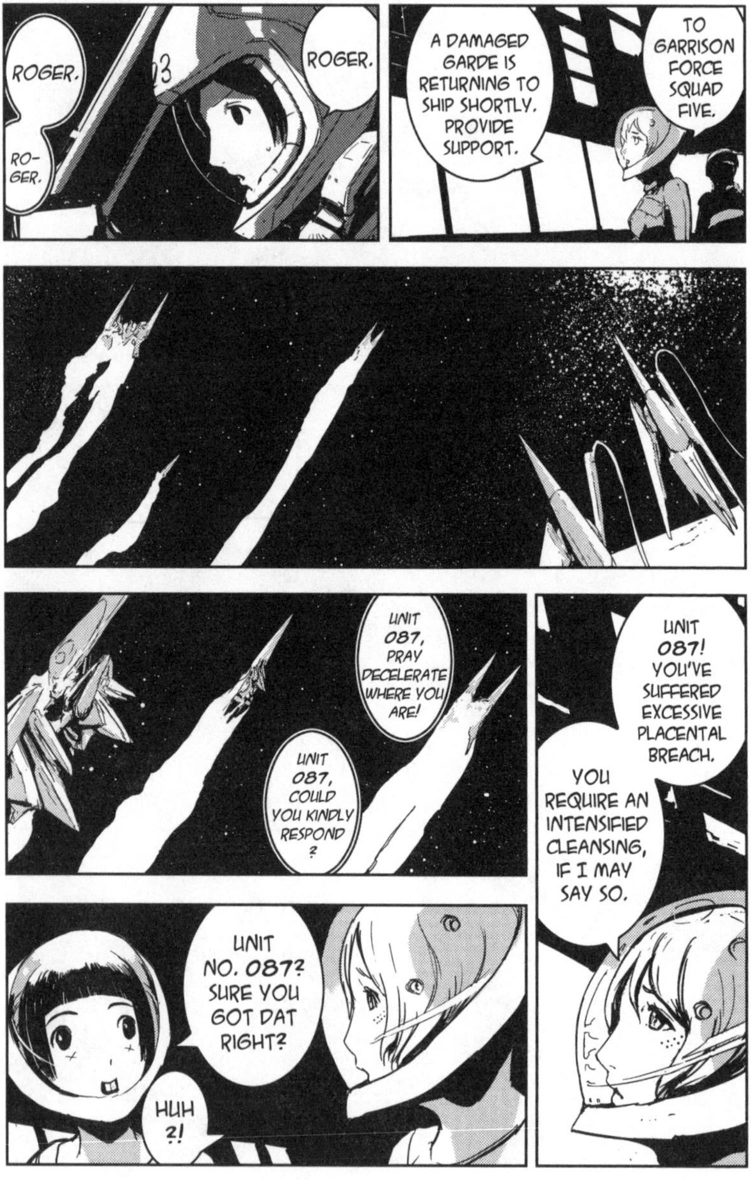
TO GARRISON FORCE SQUAD FIVE.
A DAMAGED GARDE IS RETURNING TO SHIP SHORTLY. PROVIDE SUPPORT.
ROGER.
ROGER.
RO-GER.
UNIT 087! YOU'VE SUFFERED EXCESSIVE PLACENTAL BREACH.
YOU REQUIRE AN INTENSIFIED CLEANSING, IF I MAY SAY SO.
UNIT 087, PRAY DECELERATE WHERE YOU ARE!
UNIT 087, COULD YOU KINDLY RESPOND?
UNIT NO. 087? SURE YOU GOT DAT RIGHT?
HUH?!

ユラ
WAVER

GARDE UNIT NO. 087

MAJOR DAMAGE

PILOT CONDITION UNKNOWN

UNIT 087 TOOK CRITICAL DAMAGE AN' THE PILOT BE DEAD.

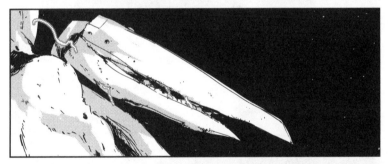

グニョ
GLLIPP

BTHOOM
BFOOO
GRRM
CHZSSK
THUD
EEK

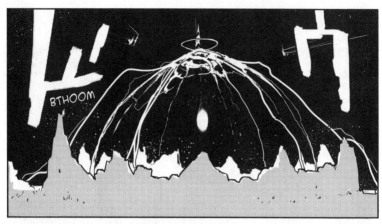

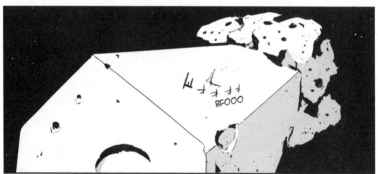

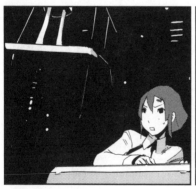

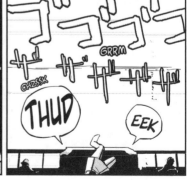

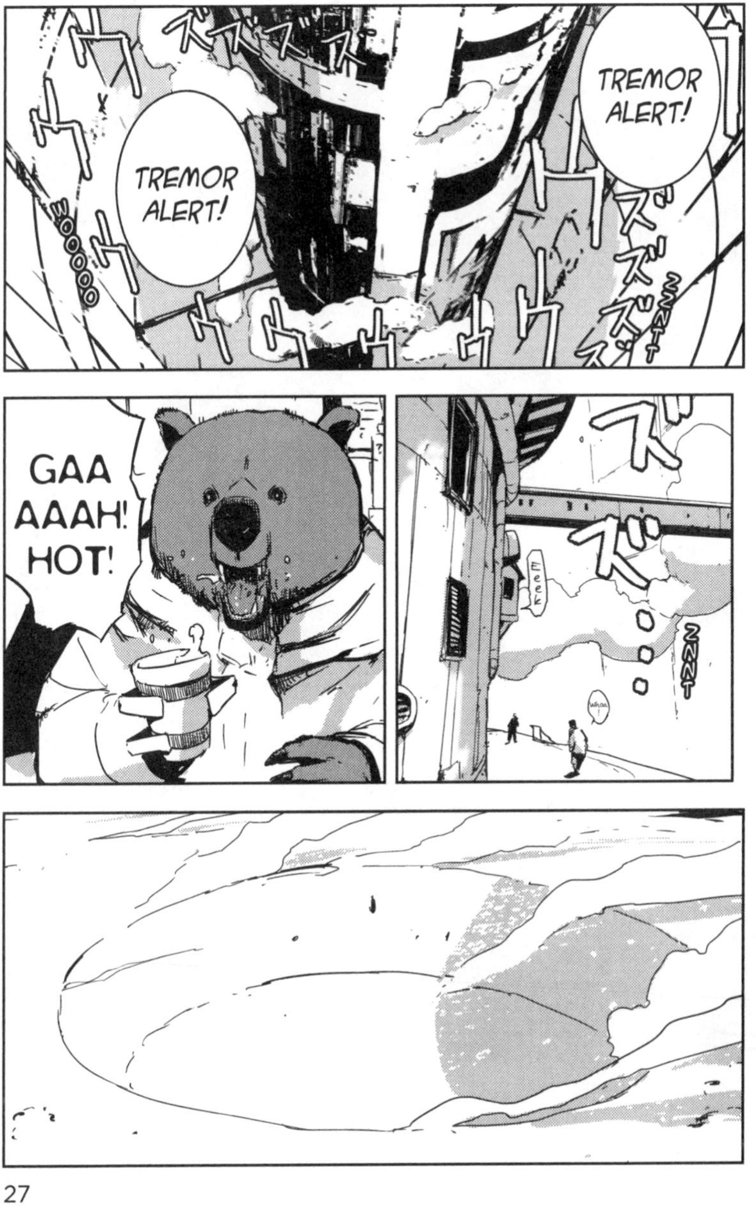
TREMOR ALERT!
ZNNT
TREMOR ALERT!
WOOOOOO
ZNNT
Eeek
Whoa!
GAA AAAH! HOT!

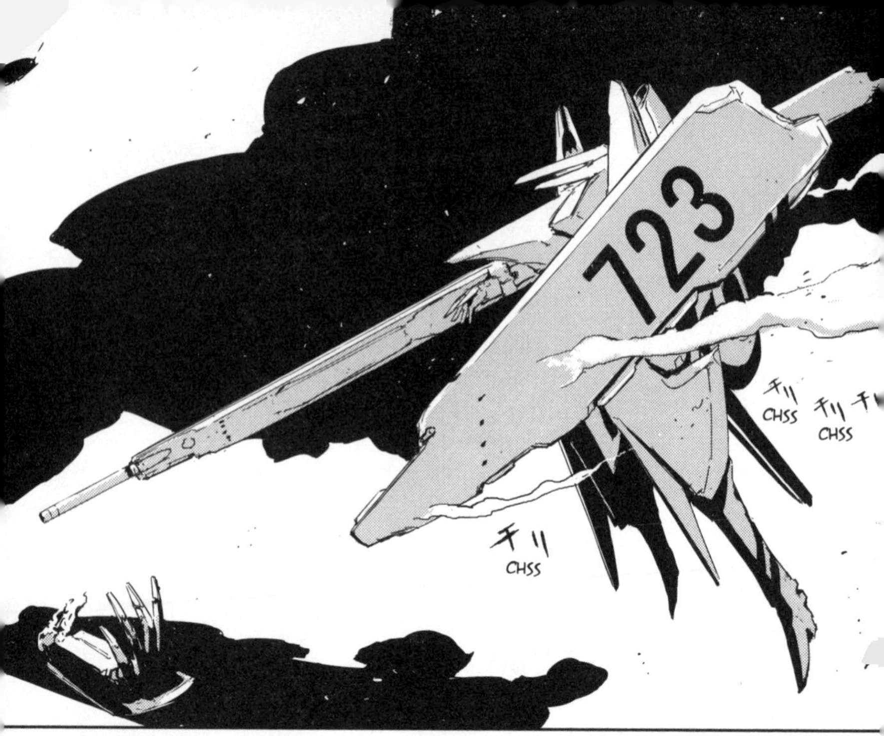

チリ CHSS
チリ CHSS
チリ CHSS

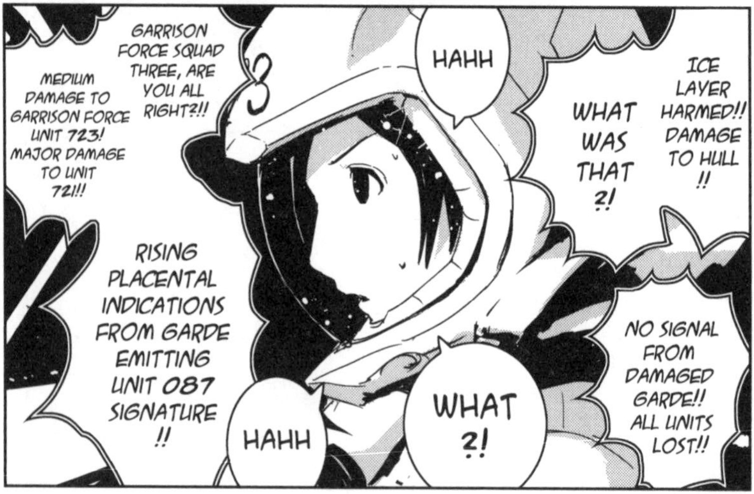

ICE LAYER HARMED!! DAMAGE TO HULL !!
WHAT WAS THAT ?!
HAHH
GARRISON FORCE SQUAD THREE, ARE YOU ALL RIGHT?!!
MEDIUM DAMAGE TO GARRISON FORCE UNIT 723! MAJOR DAMAGE TO UNIT 721!!
NO SIGNAL FROM DAMAGED GARDE!! ALL UNITS LOST!!
WHAT ?!
HAHH
RISING PLACENTAL INDICATIONS FROM GARDE EMITTING UNIT 087 SIGNATURE !!

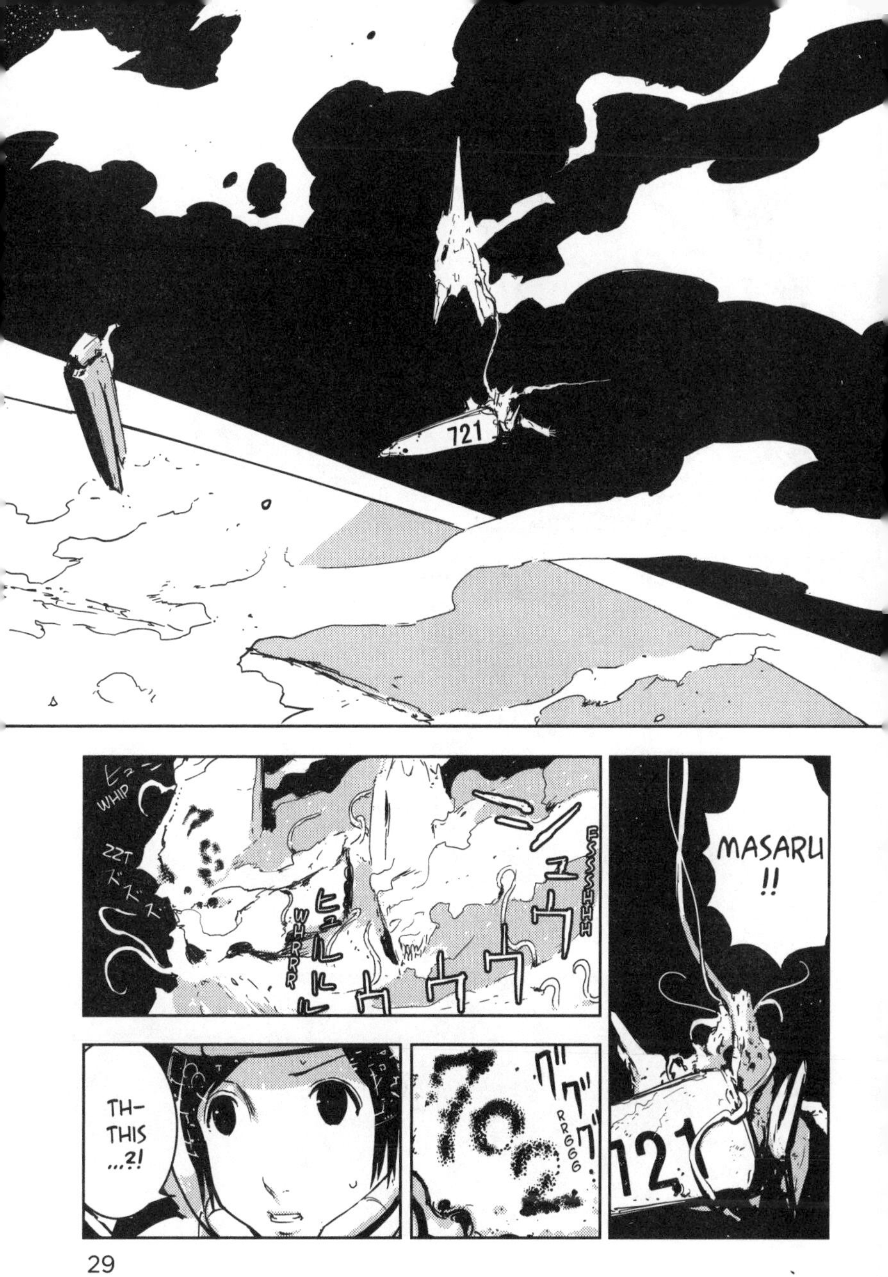
MASARU !!
FSSSSHHH
WHRRR
ZZT
WHIP
RRGGG
TH-THIS ...?!

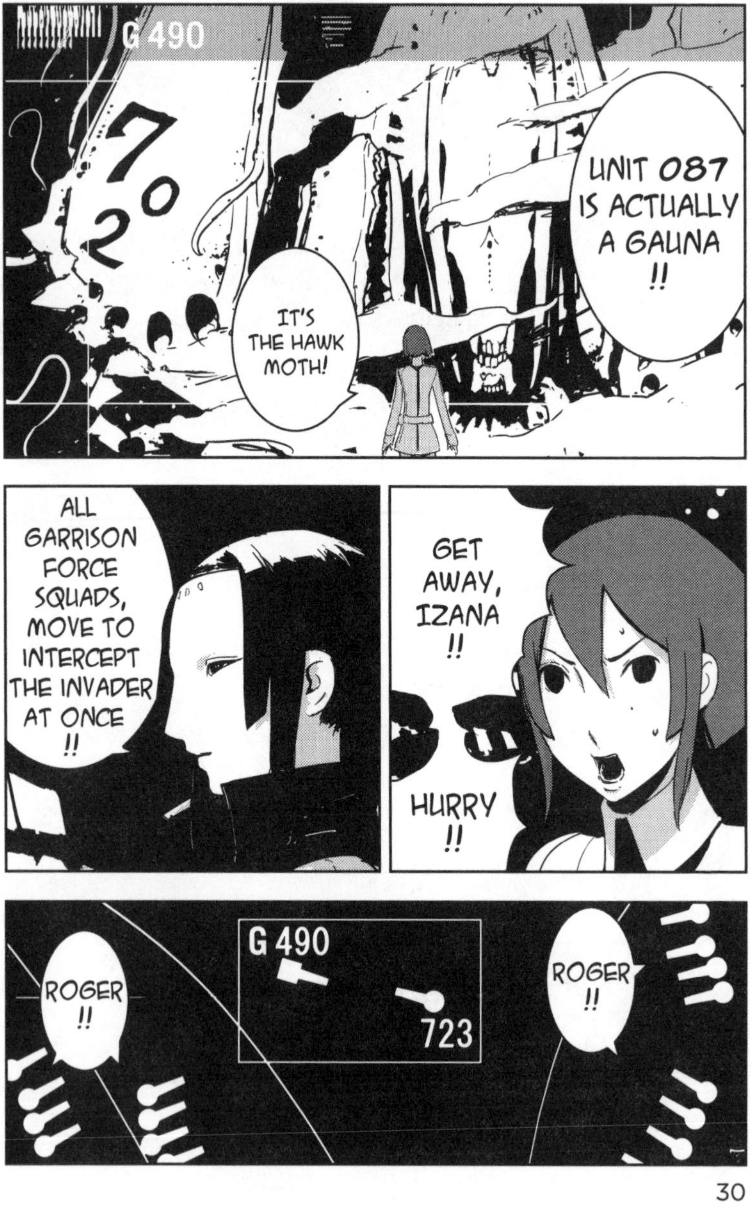
G 490
UNIT 087 IS ACTUALLY A GAUNA !!
IT'S THE HAWK MOTH!
GET AWAY, IZANA !!
HURRY !!
ALL GARRISON FORCE SQUADS, MOVE TO INTERCEPT THE INVADER AT ONCE !!
ROGER !!
G 490
723
ROGER !!

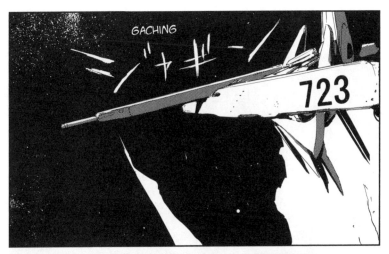

GACHING

723

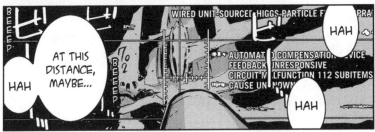

WIRED UNIT-SOURCED HIGGS-PARTICLE F... PRA...

BEEEP

AT THIS DISTANCE, MAYBE...

HAH

HAH

AUTOMAT D COMPENSATIO VICE
FEEDBACK NRESPONSIVE
CIRCUIT M LFUNCTION 112 SUBITEMS
CAUSE UN NOWN

HAH

KCHIK

IZANA !!

723

HAH

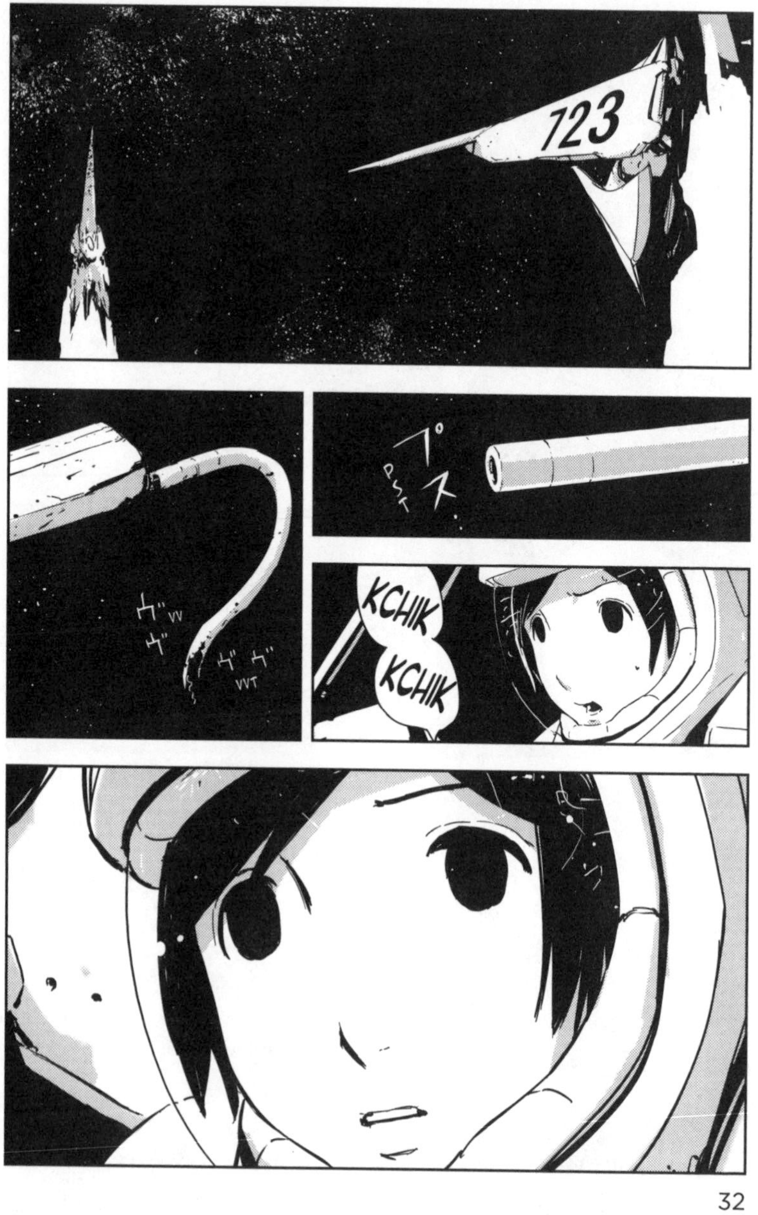
723
プス
PST
KCHIK
KCHIK
ヴ VV
ヴ
ヴヴ
VVT

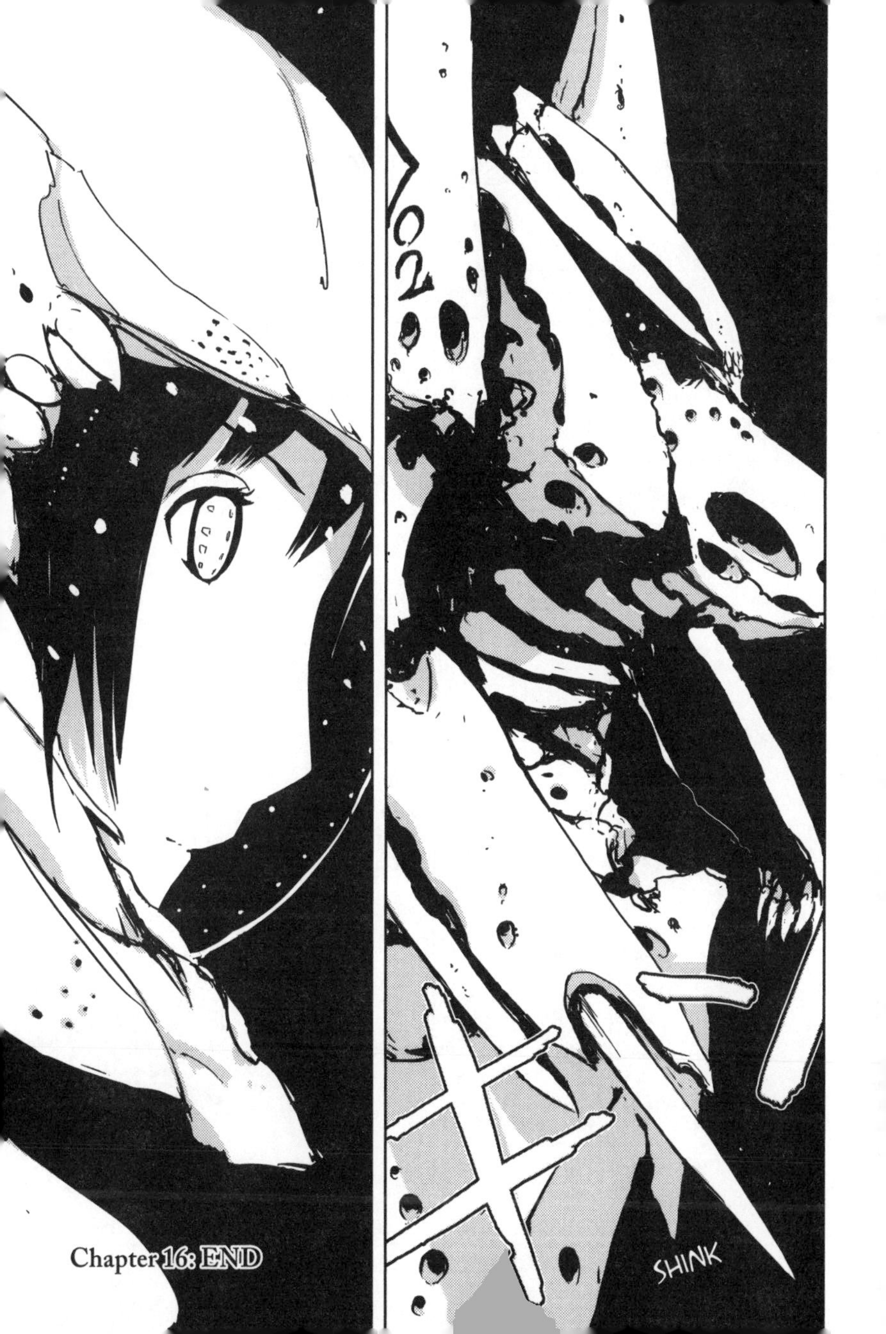
SHINK
Chapter 16: END

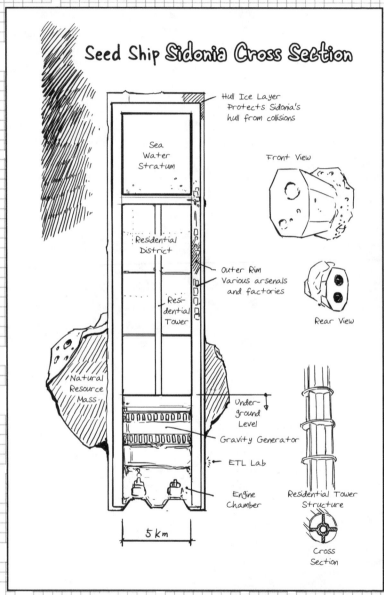

Seed Ship Sidonia Cross Section

Hull Ice Layer
Protects Sidonia's
hull from collisions

Sea Water Stratum

Front View

Residential District

Outer Rim
Various arsenals
and factories

Residential Tower

Rear View

Natural Resource Mass

Underground Level

Gravity Generator

ETL Lab

Engine Chamber

Residential Tower Structure

Cross Section

5 km

Knights of Sidonia Background Notes

Chapter 17: Nagate Tanikaze's Resolve

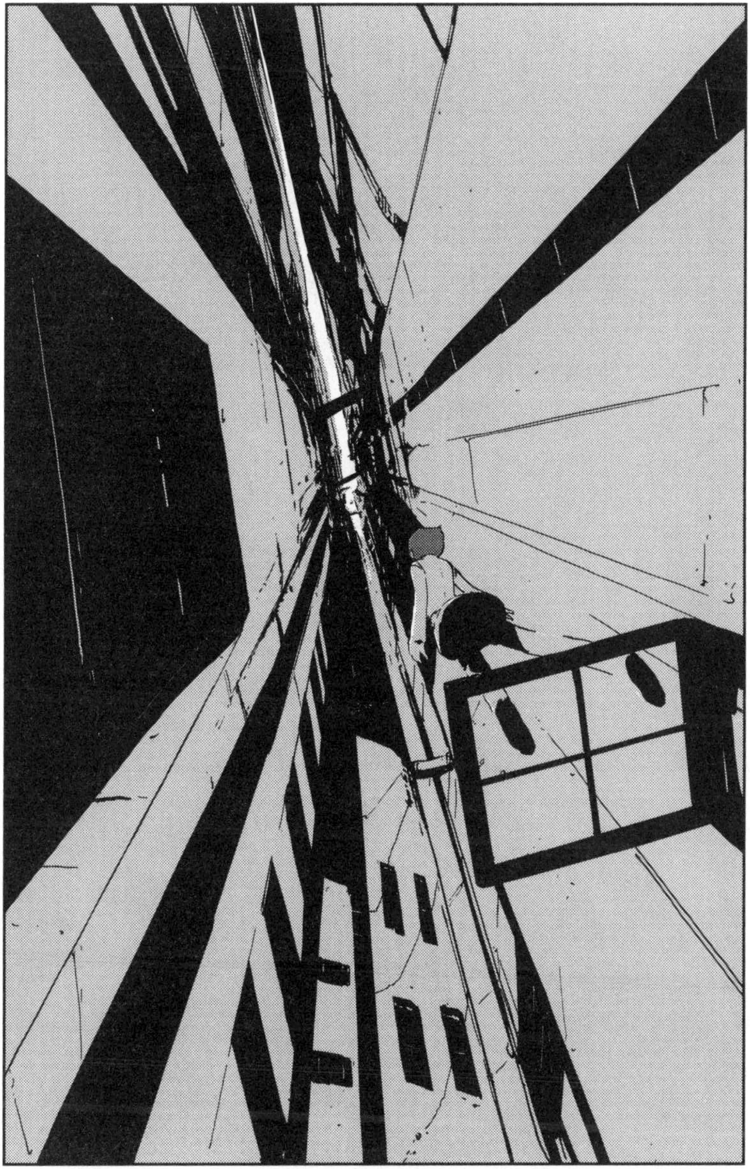

One Hundred Sights of Sidonia Part Thirteen:
Residential Tower Center Section Environs

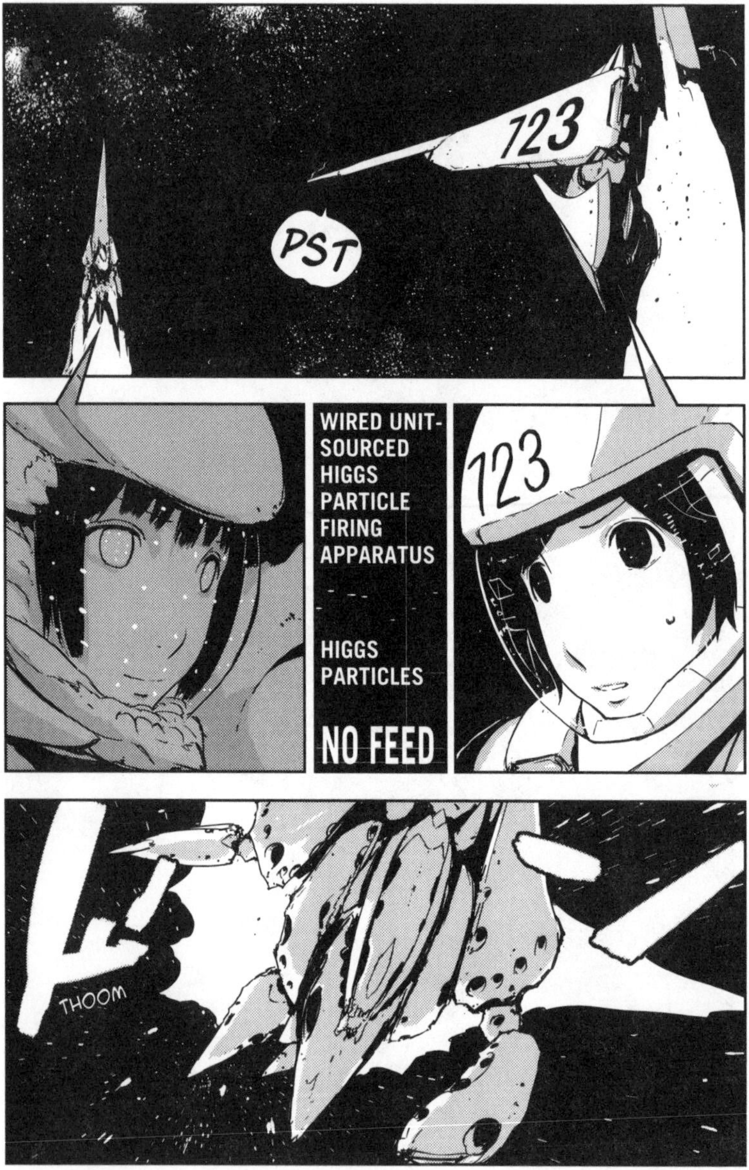
723
PST
723
WIRED UNIT-SOURCED HIGGS PARTICLE FIRING APPARATUS
HIGGS PARTICLES
NO FEED
THOOM

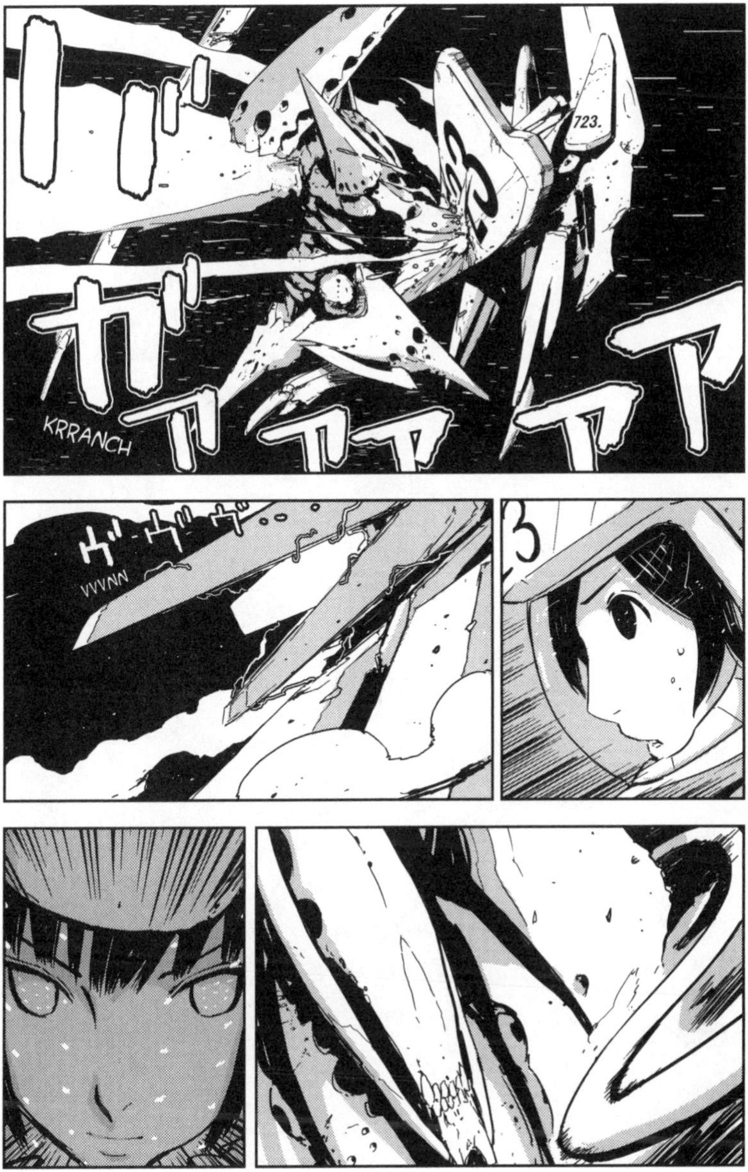
723.
KRRANCH
3
VVVNN

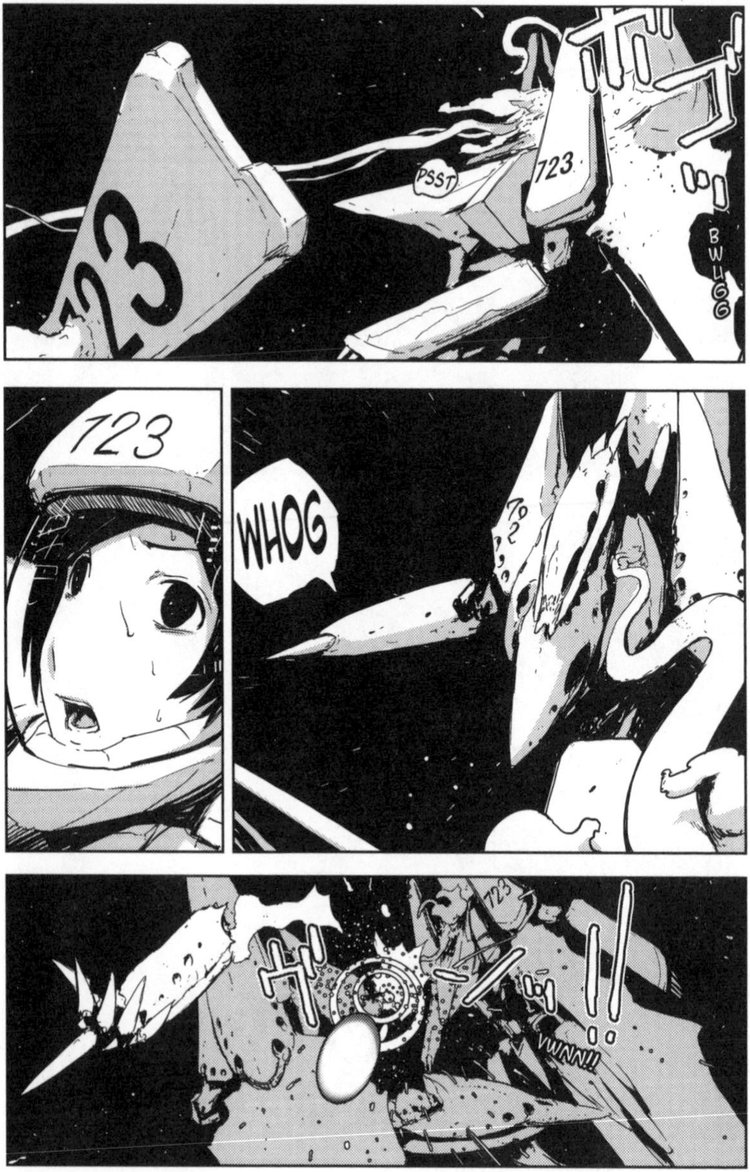
BWUGG
PSST
723
WHOG
VWNN!!

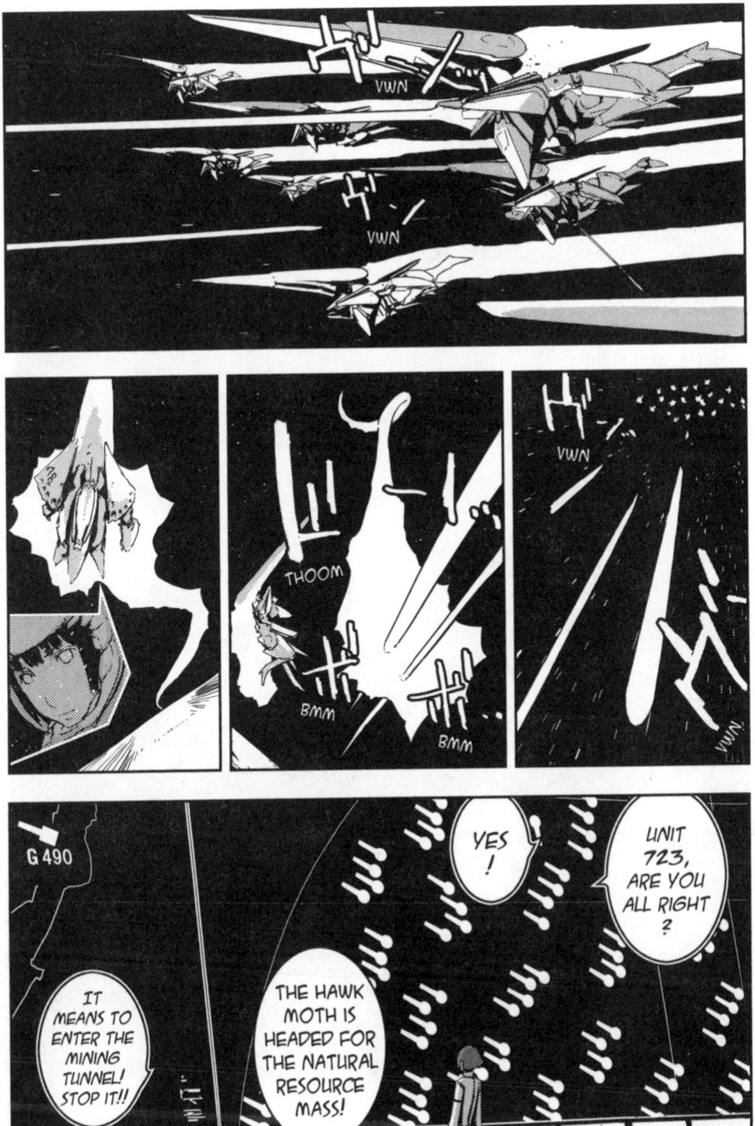
VWN
VWN
VWN
VWN
THOOM
BMM
BMM
UNIT 723, ARE YOU ALL RIGHT?
YES!
THE HAWK MOTH IS HEADED FOR THE NATURAL RESOURCE MASS!
IT MEANS TO ENTER THE MINING TUNNEL! STOP IT!!
G 490

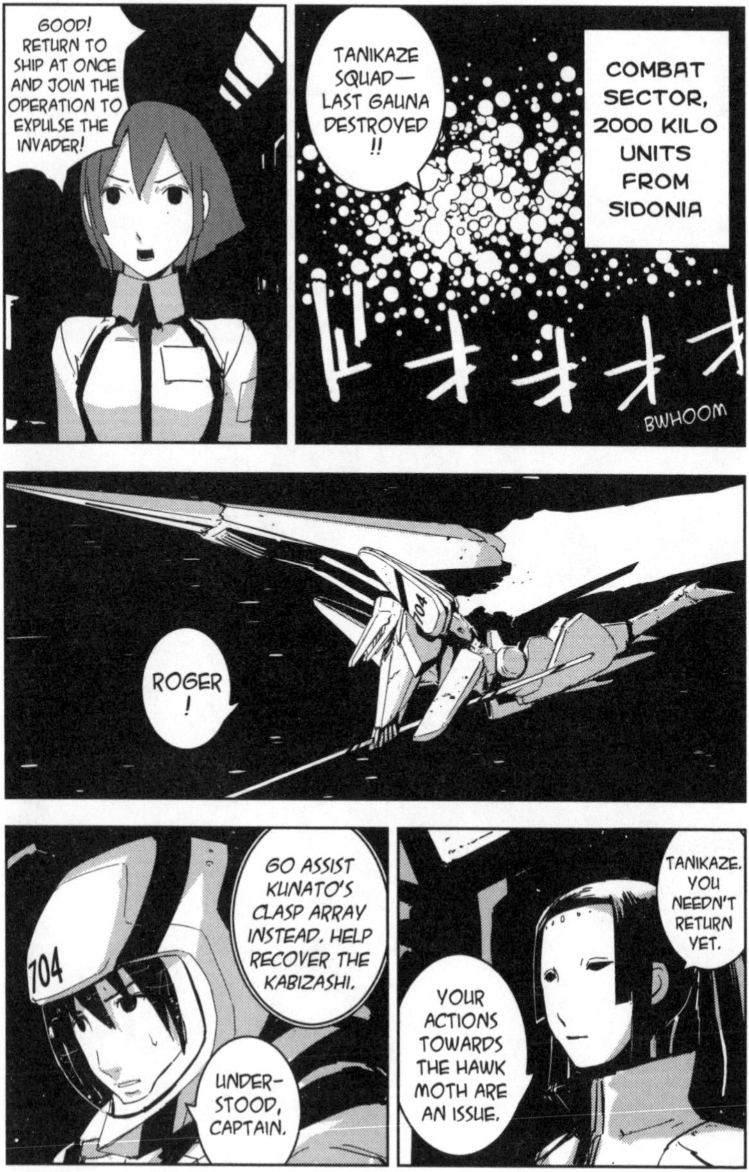
COMBAT SECTOR, 2000 KILO UNITS FROM SIDONIA
TANIKAZE SQUAD— LAST GAUNA DESTROYED !!
ドォォォォ
BWHOOM
GOOD! RETURN TO SHIP AT ONCE AND JOIN THE OPERATION TO EXPULSE THE INVADER!
ROGER !
TANIKAZE, YOU NEEDN'T RETURN YET.
YOUR ACTIONS TOWARDS THE HAWK MOTH ARE AN ISSUE.
GO ASSIST KUNATO'S CLASP ARRAY INSTEAD. HELP RECOVER THE KABIZASHI.
UNDER-STOOD, CAPTAIN.

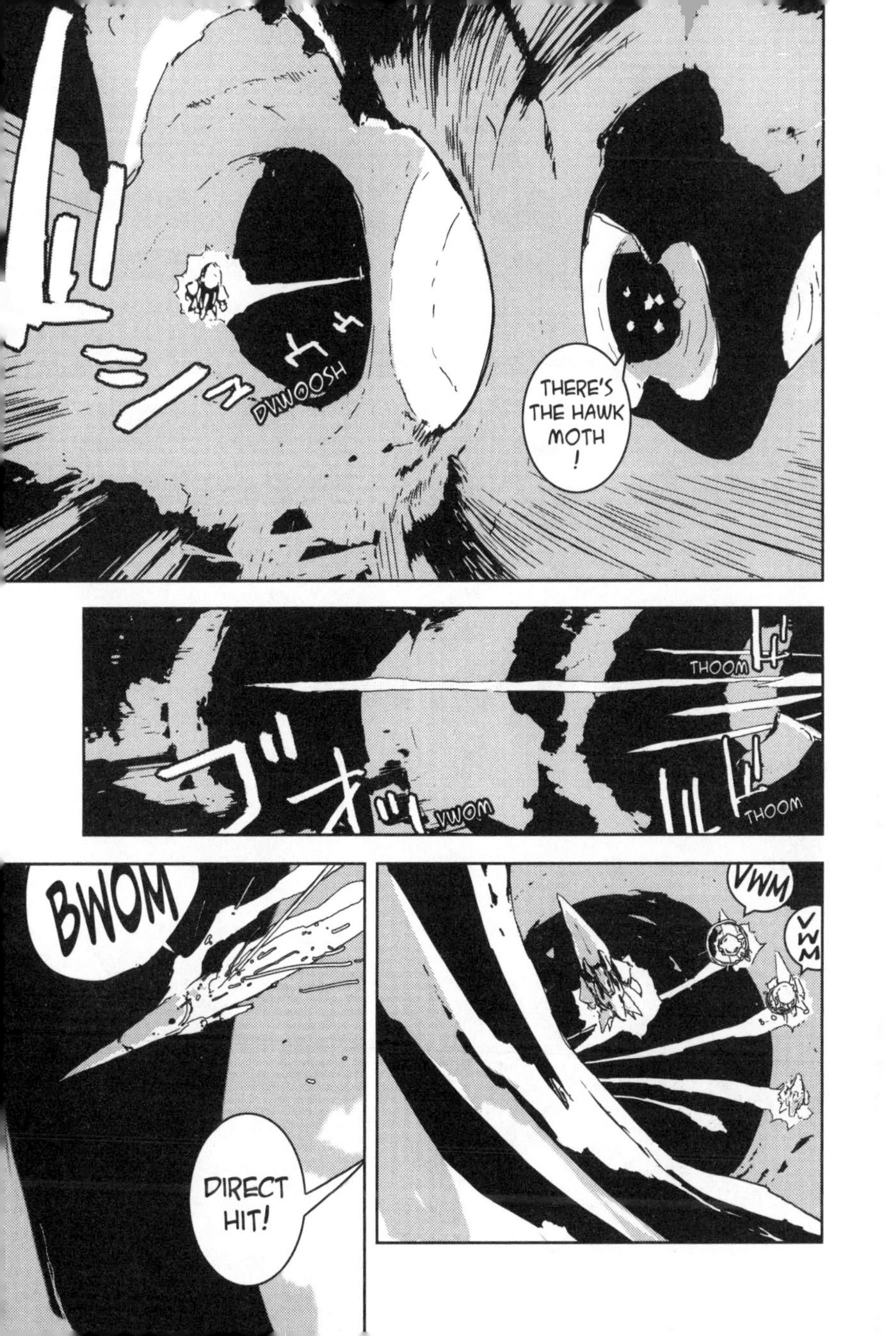
THERE'S THE HAWK MOTH!
DWOOSH
THOOM
THOOM
VWOM
VWM
VWM
BWOM
DIRECT HIT!

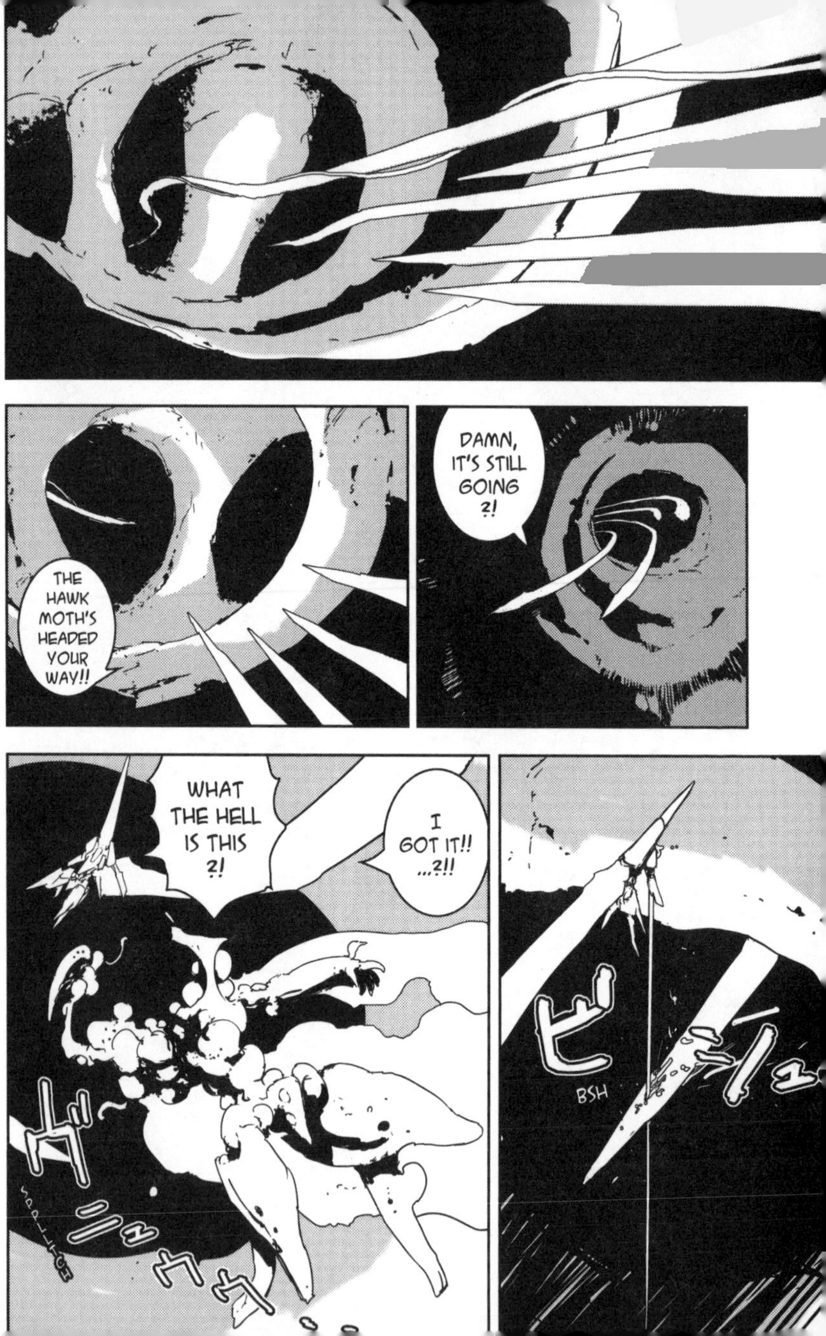
DAMN, IT'S STILL GOING ?!
THE HAWK MOTH'S HEADED YOUR WAY!!
BSH
I GOT IT!! ...?!!
WHAT THE HELL IS THIS ?!
SPPLLTCH

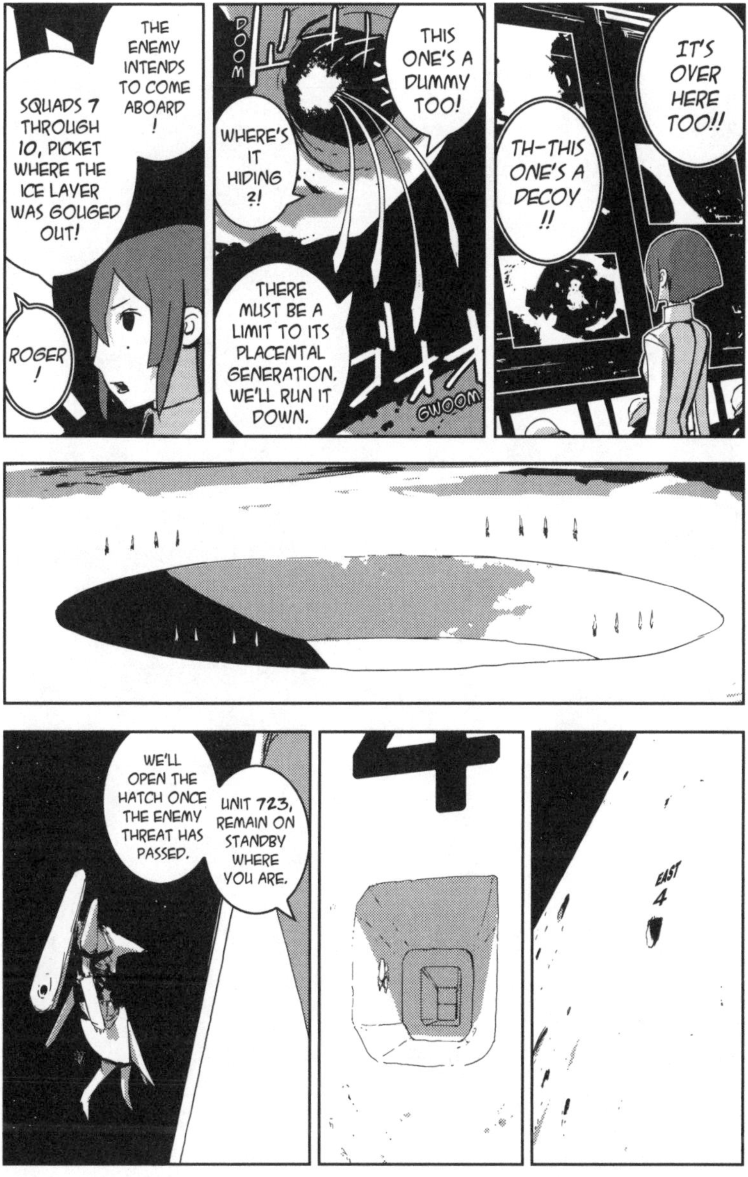
IT'S OVER HERE TOO!!
TH-THIS ONE'S A DECOY!!
DOOOM
THIS ONE'S A DUMMY TOO!
WHERE'S IT HIDING?!
THERE MUST BE A LIMIT TO ITS PLACENTAL GENERATION. WE'LL RUN IT DOWN.
GWOOM
THE ENEMY INTENDS TO COME ABOARD!
SQUADS 7 THROUGH 10, PICKET WHERE THE ICE LAYER WAS GOUGED OUT!
ROGER!
EAST 4
UNIT 723, REMAIN ON STANDBY WHERE YOU ARE.
WE'LL OPEN THE HATCH ONCE THE ENEMY THREAT HAS PASSED.

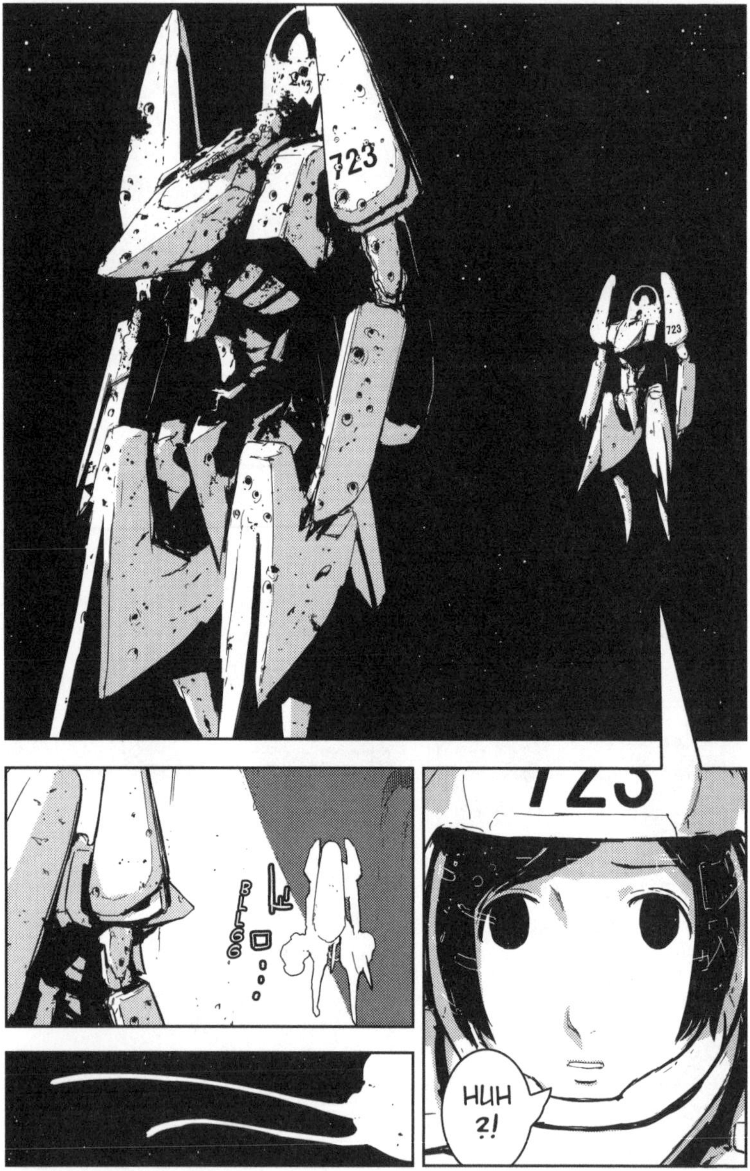
723
HUH ?!
BLLGG

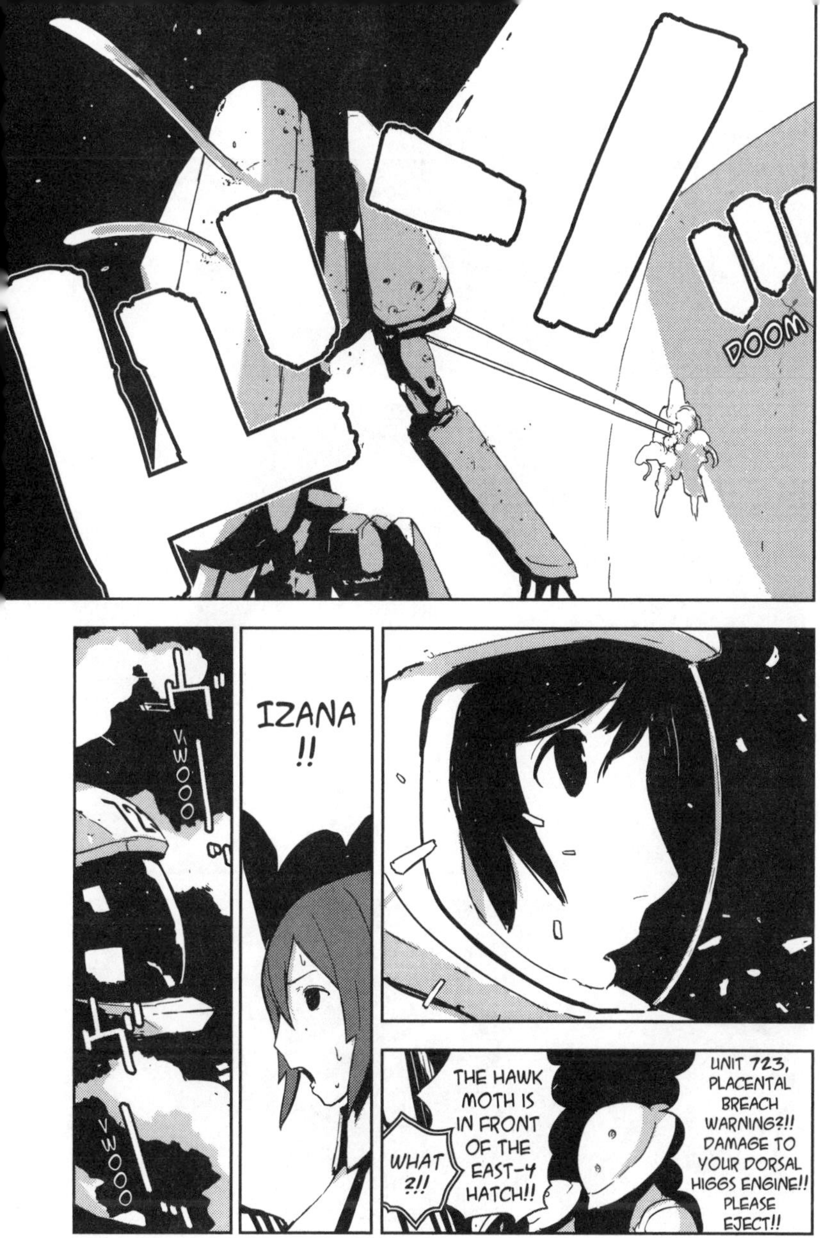
DOOM
UNIT 723, PLACENTAL BREACH WARNING?!! DAMAGE TO YOUR DORSAL HIGGS ENGINE!! PLEASE EJECT!!
THE HAWK MOTH IS IN FRONT OF THE EAST-4 HATCH!!
WHAT ?!!
IZANA !!
VWOOO
VWOOO

BLGG
UK
HAH
MANDATORY PAINKILLER ADMINISTRATION
NA...
GA...
MANDATORY PAINKIL ADMINISTRATION
FLASH

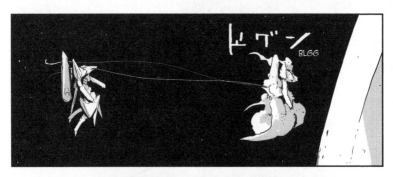

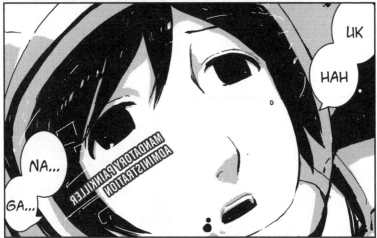

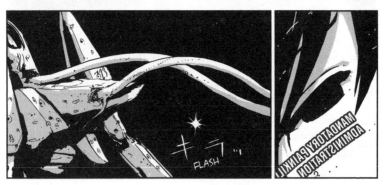

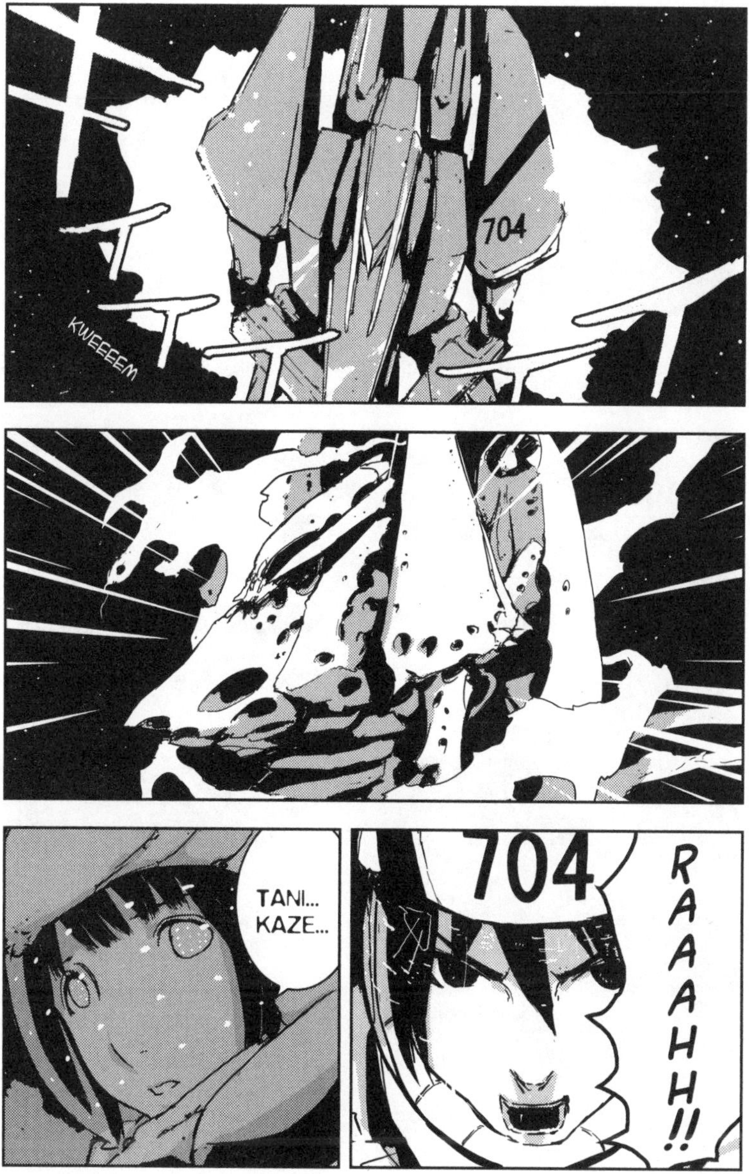
704
KWEEEEM
704
RAAAAHH!!
TANI... KAZE...

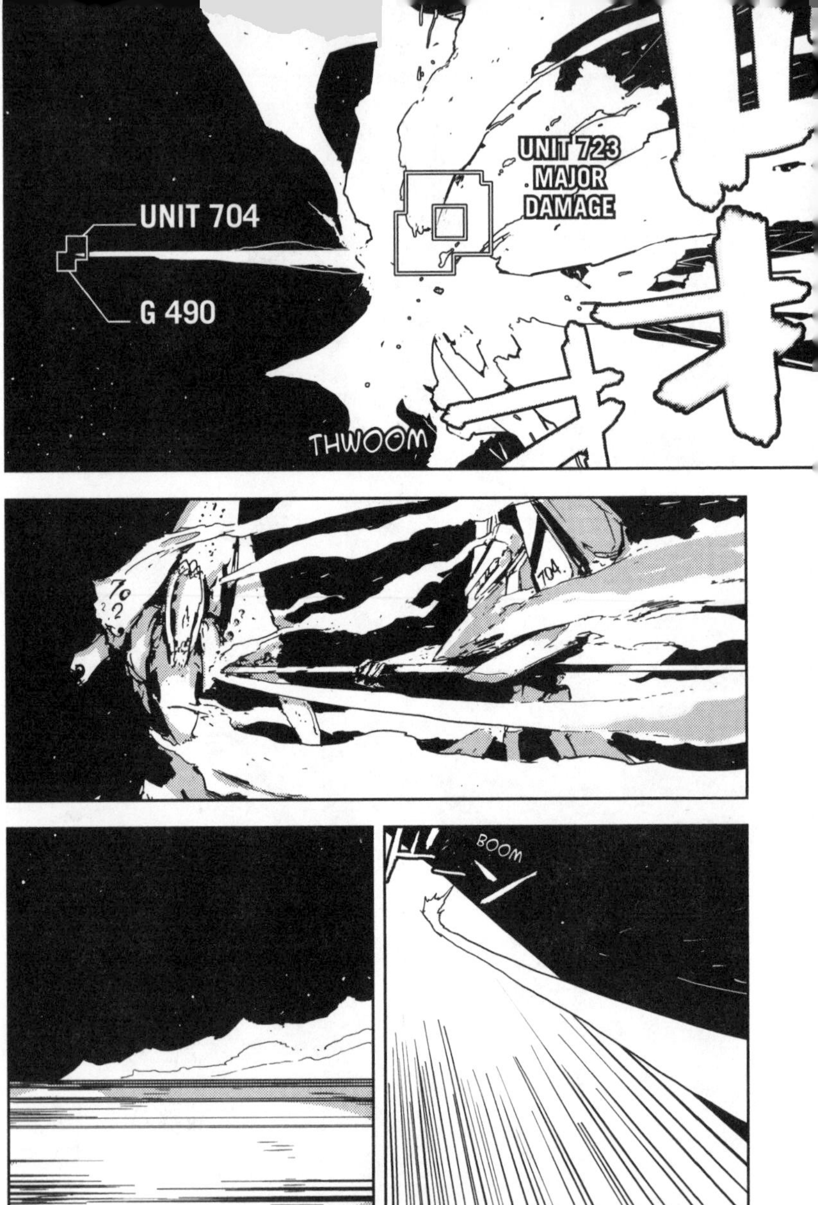
UNIT 723
MAJOR
DAMAGE
UNIT 704
G 490
THWOOM
702
704
BOOM

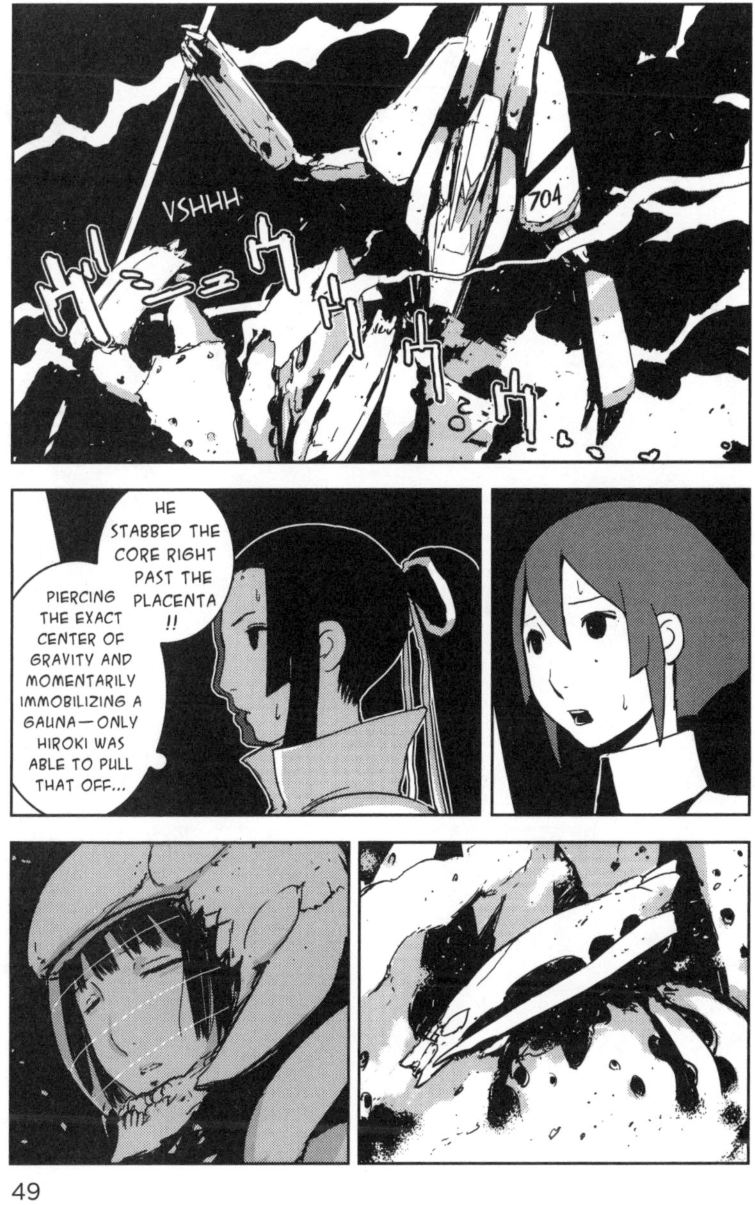
VSHHH
704
HE STABBED THE CORE RIGHT PAST THE PLACENTA !!
PIERCING THE EXACT CENTER OF GRAVITY AND MOMENTARILY IMMOBILIZING A GAUNA—ONLY HIROKI WAS ABLE TO PULL THAT OFF...

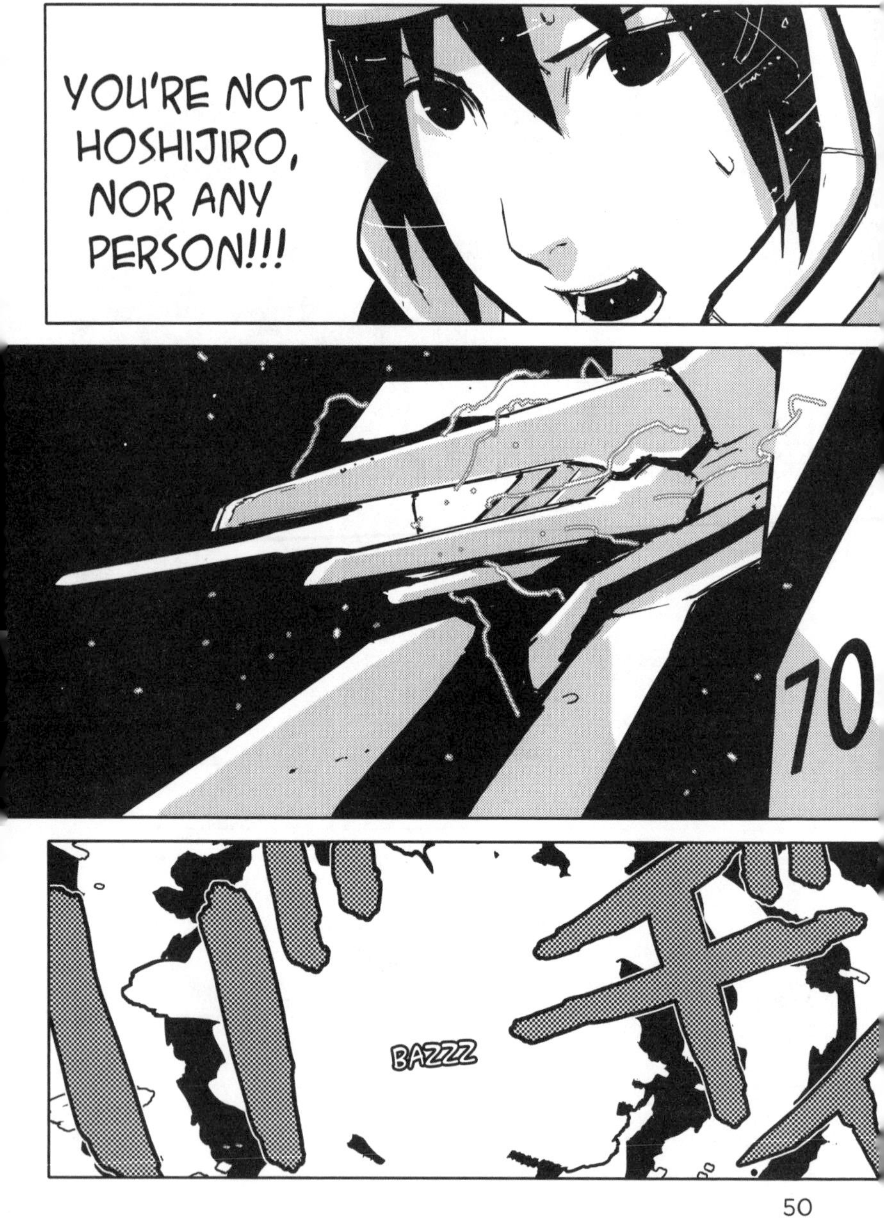
YOU'RE NOT HOSHIJIRO, NOR ANY PERSON!!!
70
BAZZZ

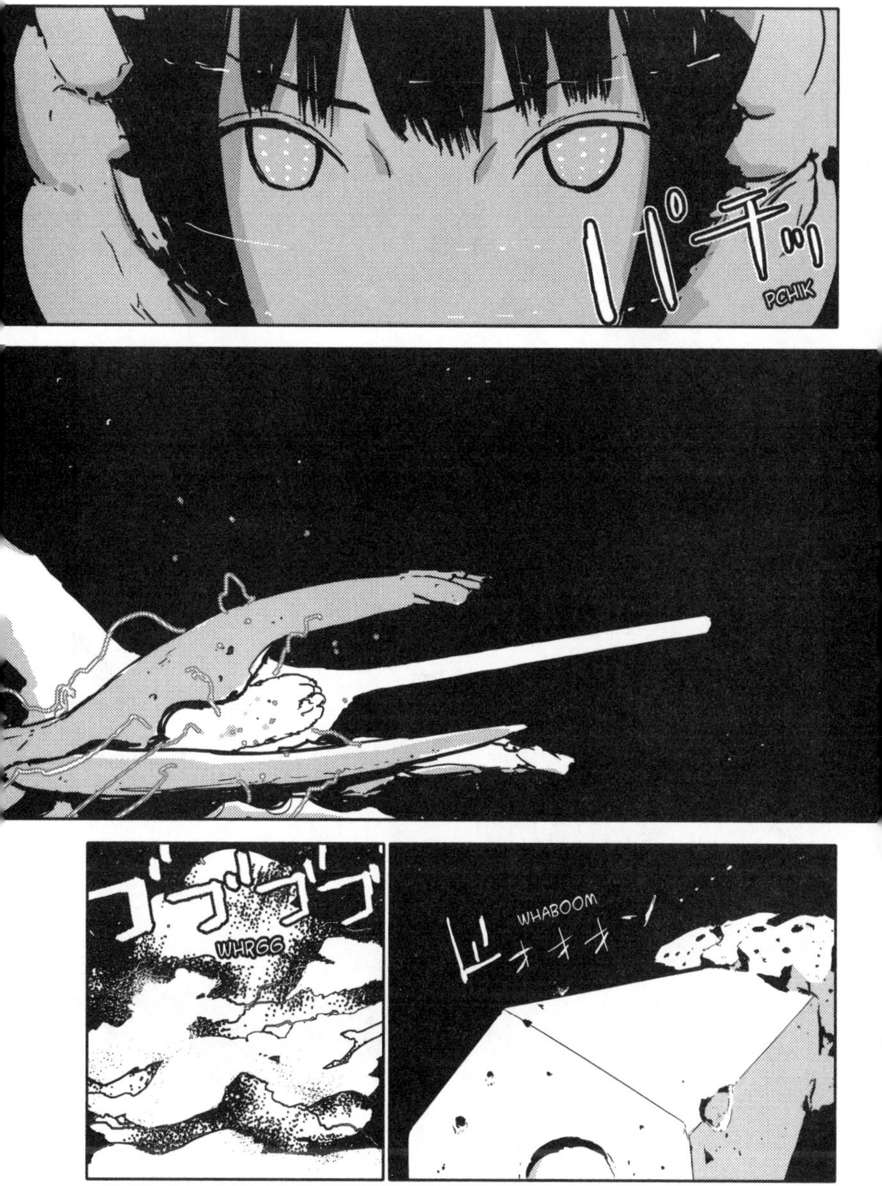
パチッ
PCHIK
WHABOOM
ゴゴゴゴ
WHRGG

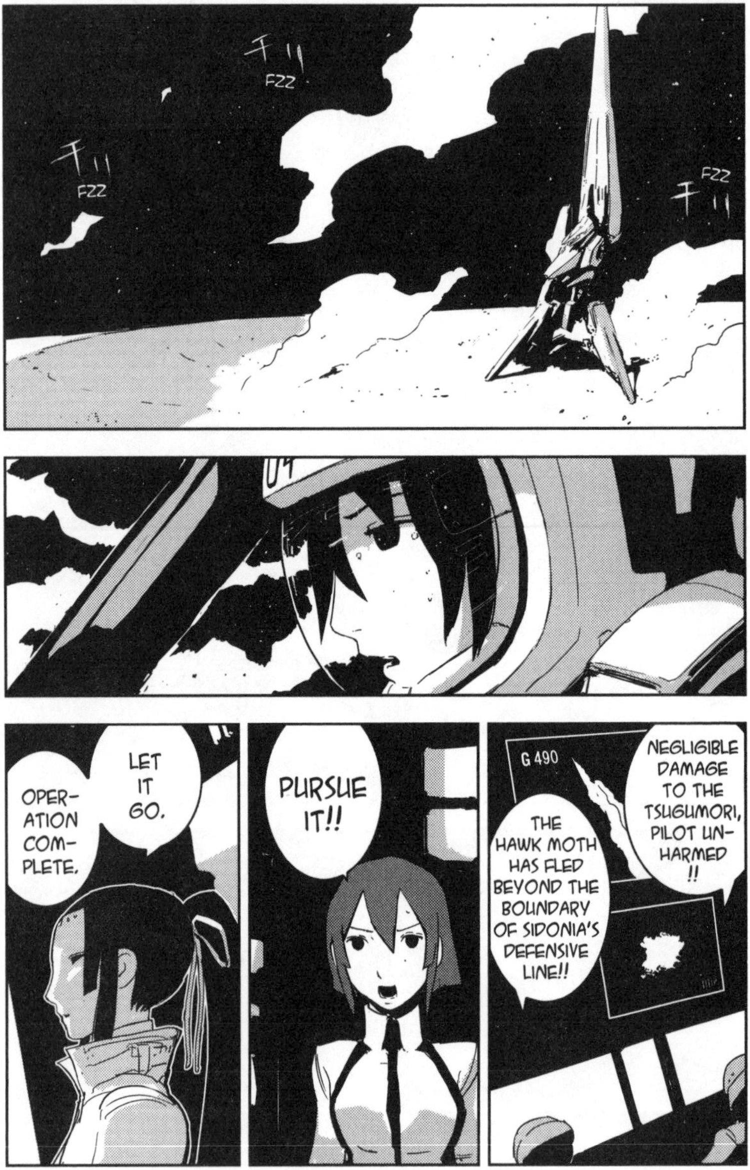
チリ
FZZ
チリ
FZZ
チリ
FZZ
G 490
NEGLIGIBLE DAMAGE TO THE TSUGUMORI, PILOT UNHARMED!!
THE HAWK MOTH HAS FLED BEYOND THE BOUNDARY OF SIDONIA'S DEFENSIVE LINE!!
PURSUE IT!!
LET IT GO.
OPERATION COMPLETE.

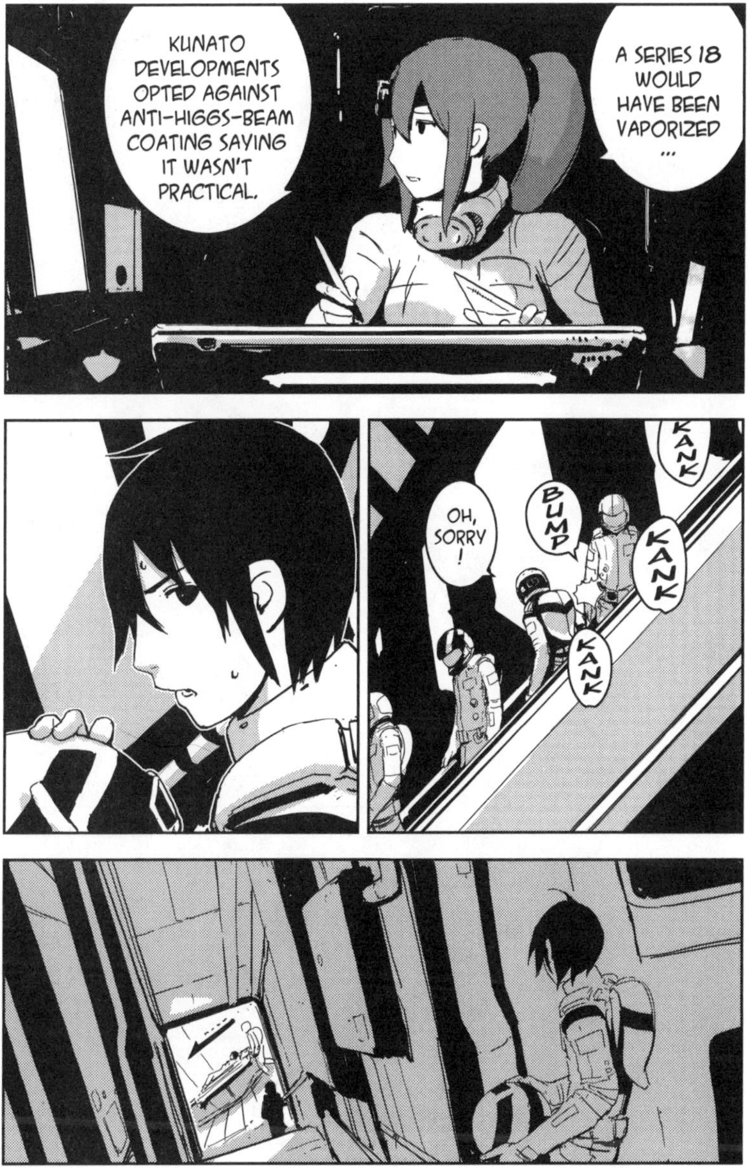
A SERIES 18 WOULD HAVE BEEN VAPORIZED ...
KUNATO DEVELOPMENTS OPTED AGAINST ANTI-HIGGS-BEAM COATING SAYING IT WASN'T PRACTICAL.
KANK
BUMP
OH, SORRY!
KANK
KANK

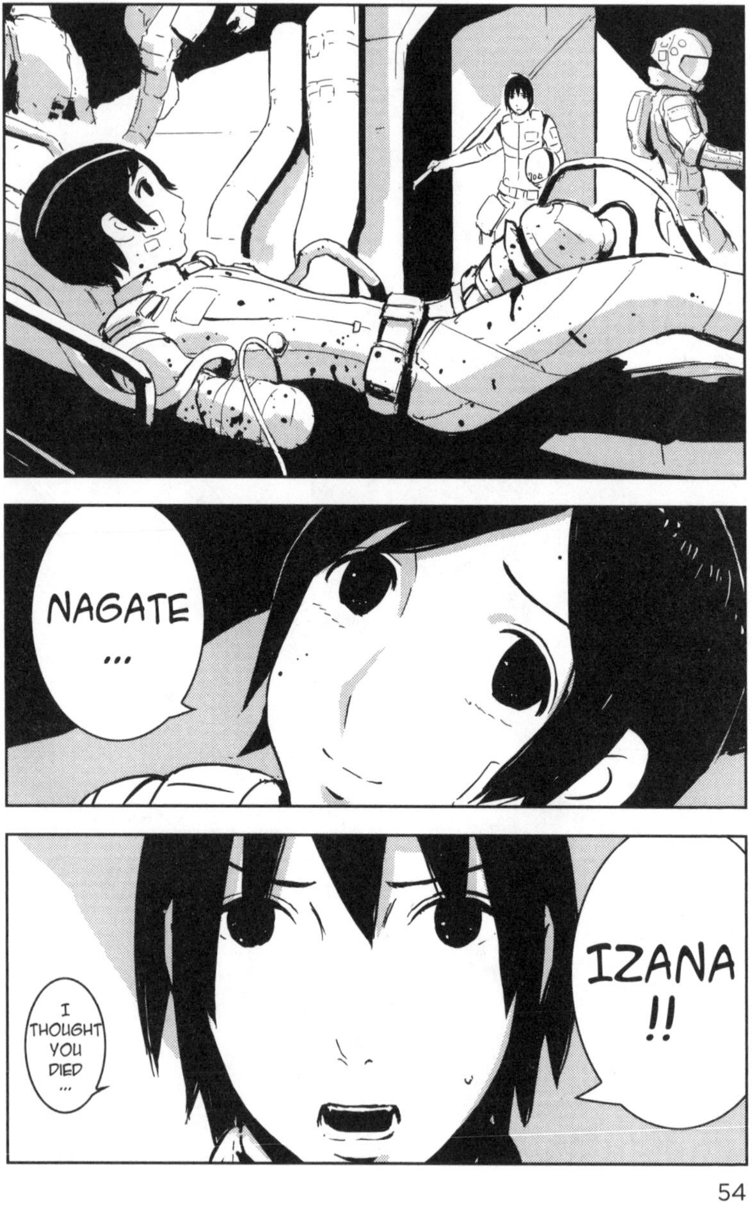
NAGATE ...
IZANA !!
I THOUGHT YOU DIED ...

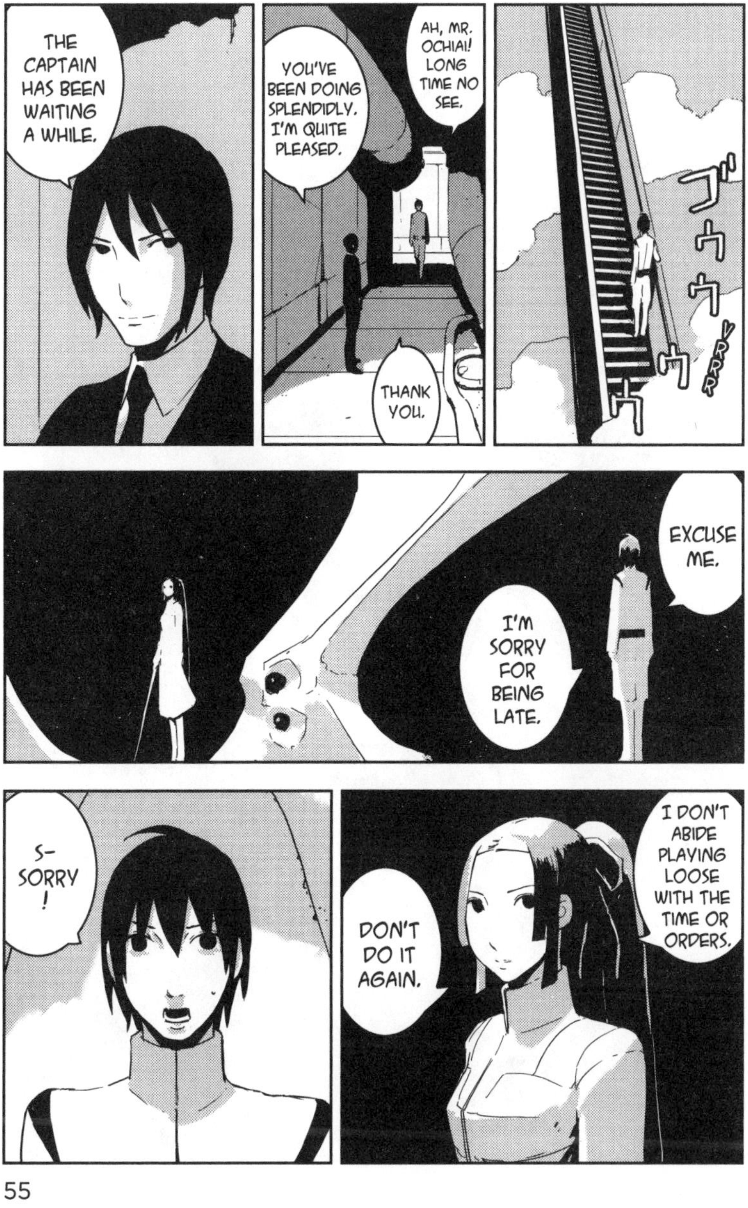
ゴウウウウ
VRRR
AH, MR. OCHIAI! LONG TIME NO SEE.
YOU'VE BEEN DOING SPLENDIDLY. I'M QUITE PLEASED.
THANK YOU.
THE CAPTAIN HAS BEEN WAITING A WHILE.
EXCUSE ME.
I'M SORRY FOR BEING LATE.
I DON'T ABIDE PLAYING LOOSE WITH THE TIME OR ORDERS.
DON'T DO IT AGAIN.
S-SORRY!

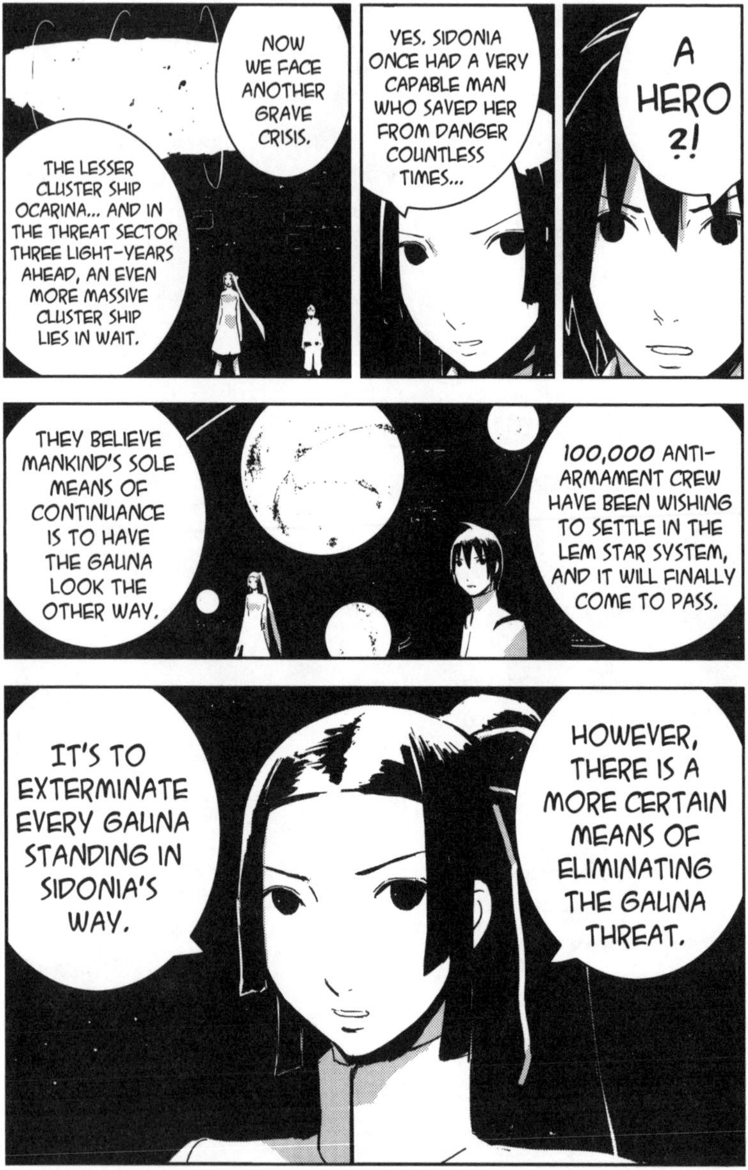
A HERO?!
YES. SIDONIA ONCE HAD A VERY CAPABLE MAN WHO SAVED HER FROM DANGER COUNTLESS TIMES...
NOW WE FACE ANOTHER GRAVE CRISIS.
THE LESSER CLUSTER SHIP OCARINA... AND IN THE THREAT SECTOR THREE LIGHT-YEARS AHEAD, AN EVEN MORE MASSIVE CLUSTER SHIP LIES IN WAIT.
100,000 ANTI-ARMAMENT CREW HAVE BEEN WISHING TO SETTLE IN THE LEM STAR SYSTEM, AND IT WILL FINALLY COME TO PASS.
THEY BELIEVE MANKIND'S SOLE MEANS OF CONTINUANCE IS TO HAVE THE GAUNA LOOK THE OTHER WAY.
HOWEVER, THERE IS A MORE CERTAIN MEANS OF ELIMINATING THE GAUNA THREAT.
IT'S TO EXTERMINATE EVERY GAUNA STANDING IN SIDONIA'S WAY.

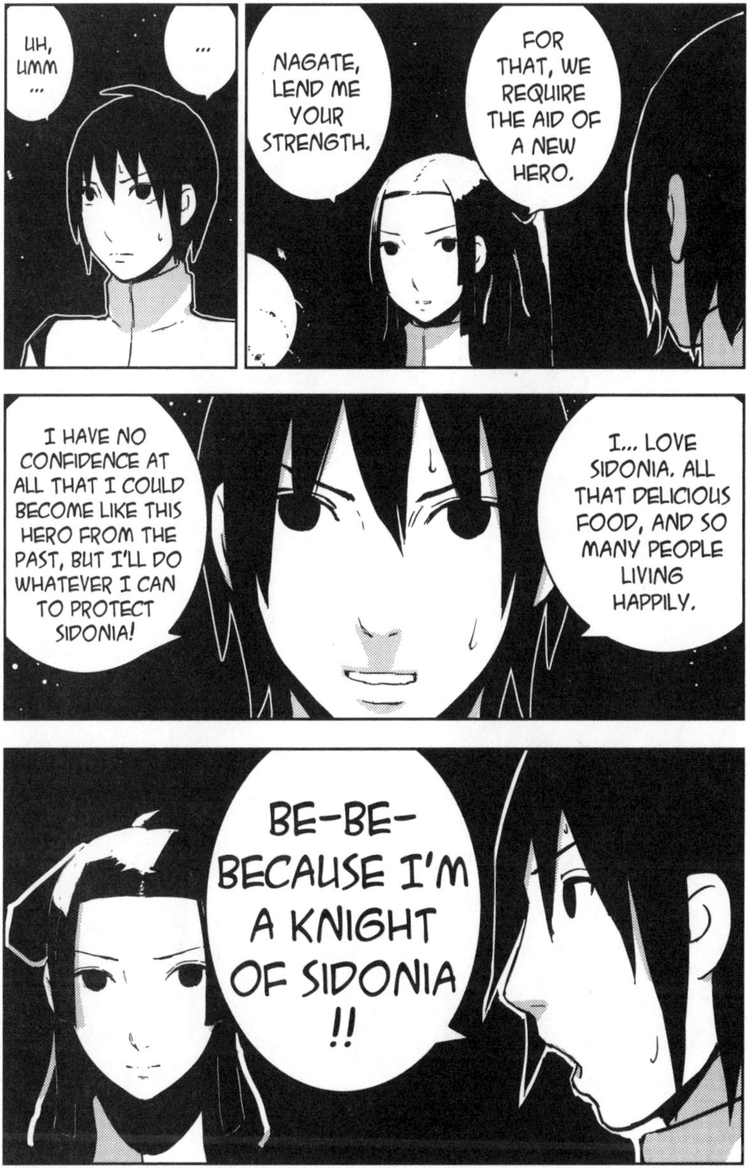
FOR THAT, WE REQUIRE THE AID OF A NEW HERO.
NAGATE, LEND ME YOUR STRENGTH.
...
UH, UMM ...
I... LOVE SIDONIA. ALL THAT DELICIOUS FOOD, AND SO MANY PEOPLE LIVING HAPPILY.
I HAVE NO CONFIDENCE AT ALL THAT I COULD BECOME LIKE THIS HERO FROM THE PAST, BUT I'LL DO WHATEVER I CAN TO PROTECT SIDONIA!
BE-BE-BECAUSE I'M A KNIGHT OF SIDONIA !!

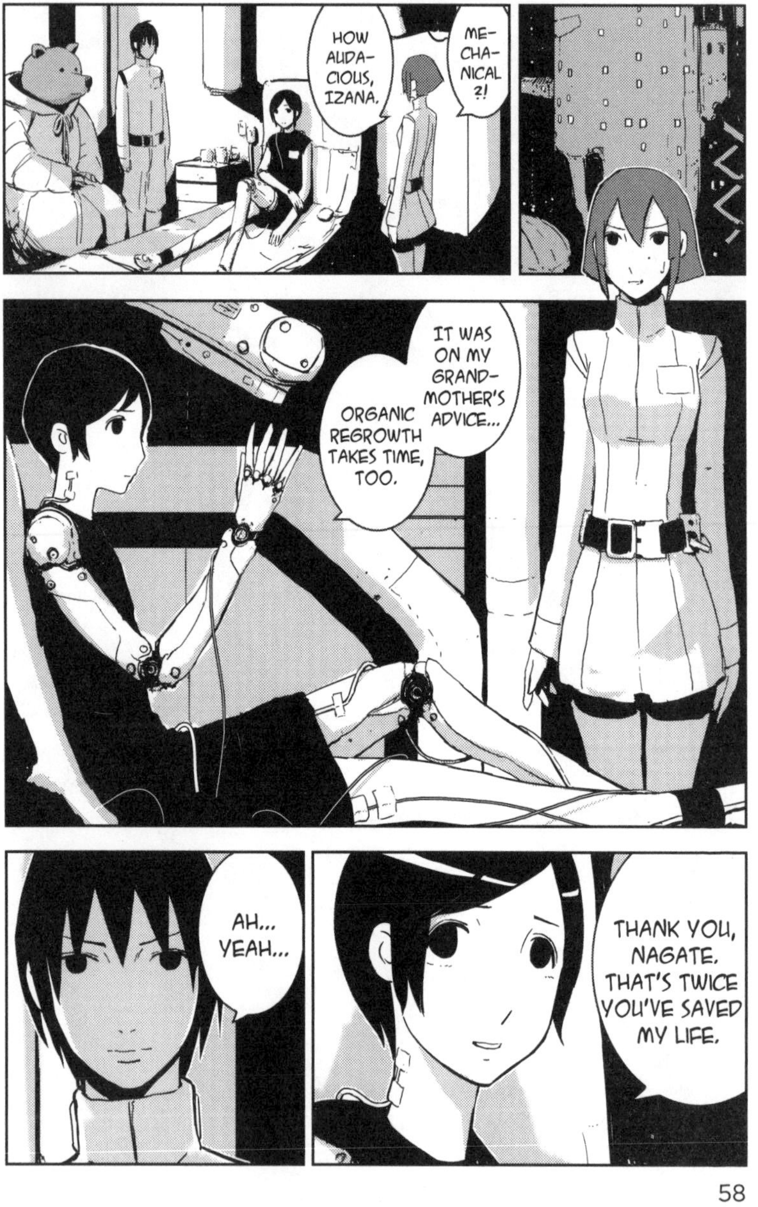
MECHANICAL?!
HOW AUDACIOUS, IZANA.
IT WAS ON MY GRANDMOTHER'S ADVICE...
ORGANIC REGROWTH TAKES TIME, TOO.
THANK YOU, NAGATE. THAT'S TWICE YOU'VE SAVED MY LIFE.
AH... YEAH...

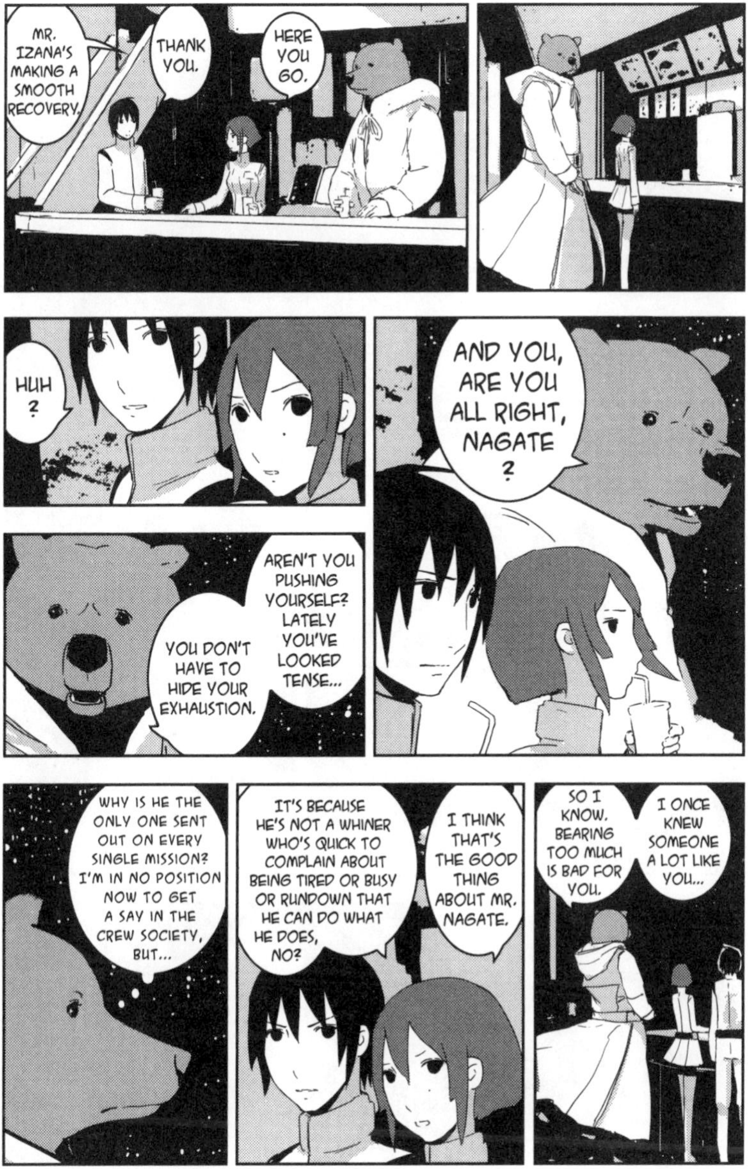
HERE YOU GO.
THANK YOU.
MR. IZANA'S MAKING A SMOOTH RECOVERY.
AND YOU, ARE YOU ALL RIGHT, NAGATE?
HUH?
AREN'T YOU PUSHING YOURSELF? LATELY YOU'VE LOOKED TENSE...
YOU DON'T HAVE TO HIDE YOUR EXHAUSTION.
I ONCE KNEW SOMEONE A LOT LIKE YOU...
SO I KNOW. BEARING TOO MUCH IS BAD FOR YOU.
I THINK THAT'S THE GOOD THING ABOUT MR. NAGATE.
IT'S BECAUSE HE'S NOT A WHINER WHO'S QUICK TO COMPLAIN ABOUT BEING TIRED OR BUSY OR RUNDOWN THAT HE CAN DO WHAT HE DOES, NO?
WHY IS HE THE ONLY ONE SENT OUT ON EVERY SINGLE MISSION? I'M IN NO POSITION NOW TO GET A SAY IN THE CREW SOCIETY, BUT...

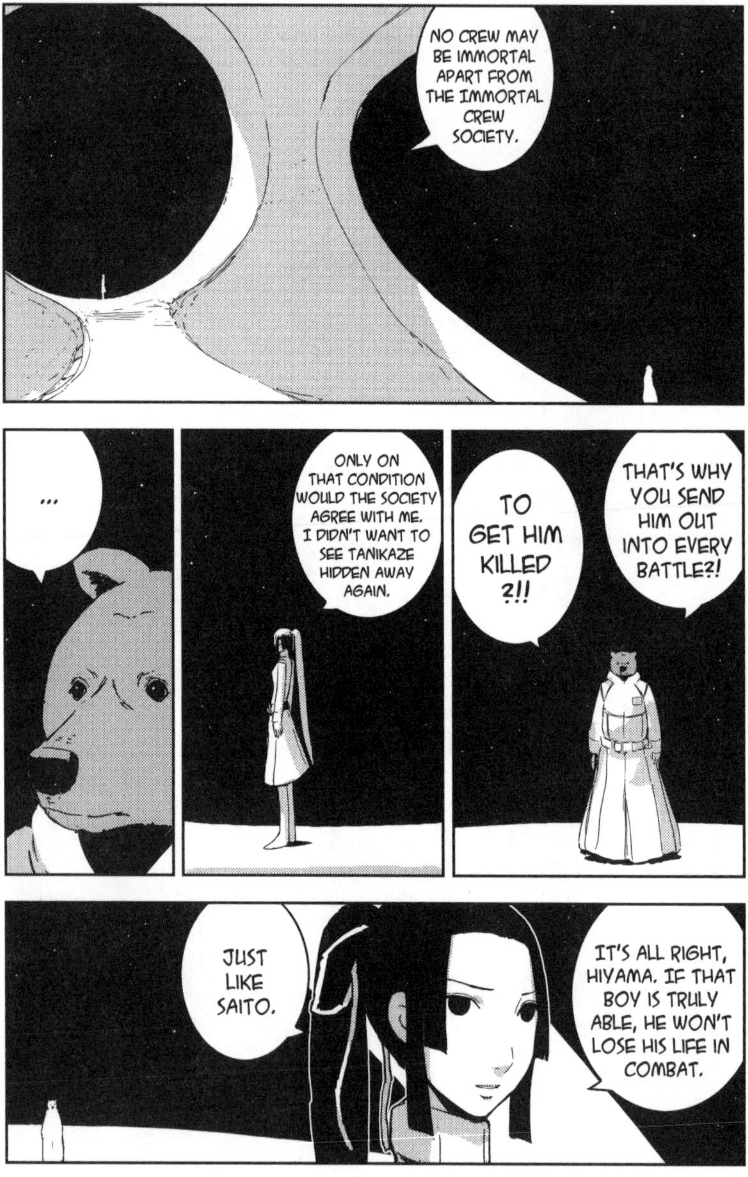
NO CREW MAY BE IMMORTAL APART FROM THE IMMORTAL CREW SOCIETY.
THAT'S WHY YOU SEND HIM OUT INTO EVERY BATTLE?!
TO GET HIM KILLED ?!!
ONLY ON THAT CONDITION WOULD THE SOCIETY AGREE WITH ME. I DIDN'T WANT TO SEE TANIKAZE HIDDEN AWAY AGAIN.
...
IT'S ALL RIGHT, HIYAMA. IF THAT BOY IS TRULY ABLE, HE WON'T LOSE HIS LIFE IN COMBAT.
JUST LIKE SAITO.

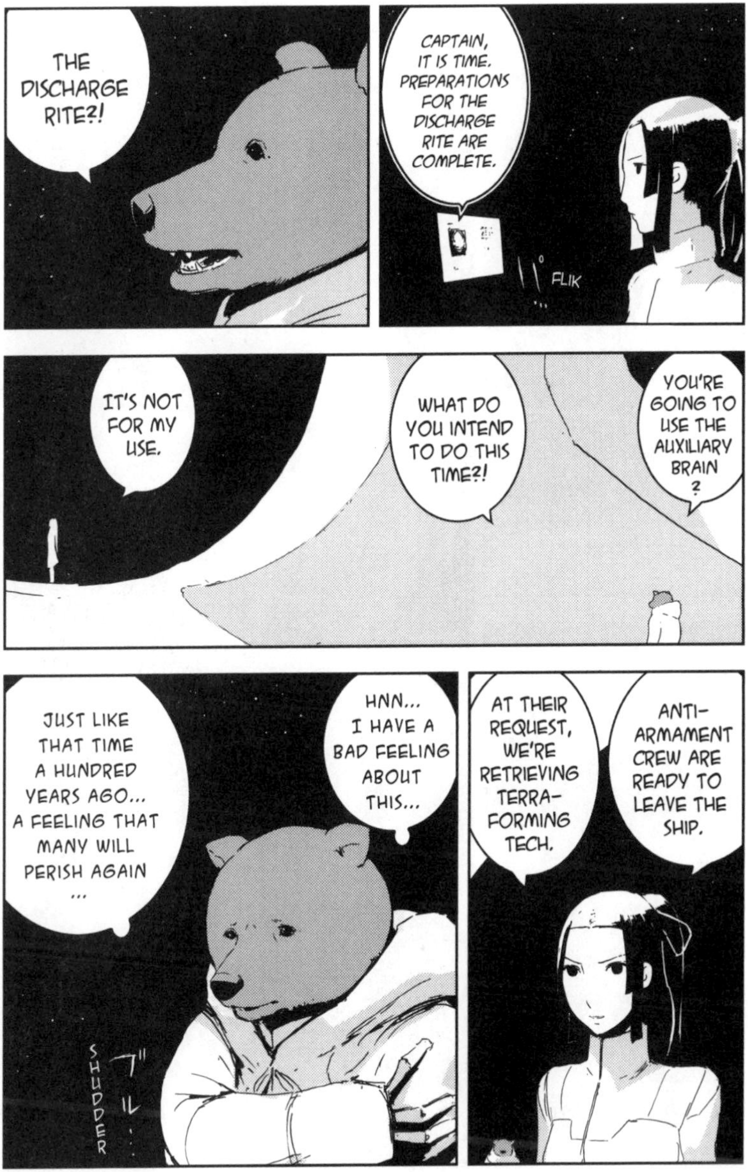
CAPTAIN, IT IS TIME. PREPARATIONS FOR THE DISCHARGE RITE ARE COMPLETE.
FLIK
THE DISCHARGE RITE?!
YOU'RE GOING TO USE THE AUXILIARY BRAIN?
WHAT DO YOU INTEND TO DO THIS TIME?!
IT'S NOT FOR MY USE.
ANTI-ARMAMENT CREW ARE READY TO LEAVE THE SHIP.
AT THEIR REQUEST, WE'RE RETRIEVING TERRA-FORMING TECH.
HNN... I HAVE A BAD FEELING ABOUT THIS...
JUST LIKE THAT TIME A HUNDRED YEARS AGO... A FEELING THAT MANY WILL PERISH AGAIN...
ブル
SHUDDER

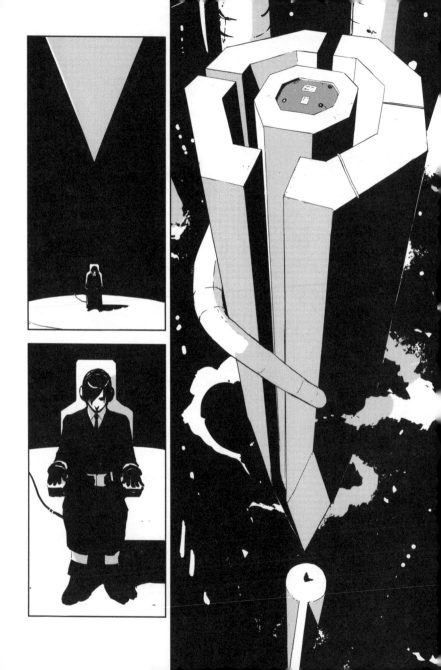

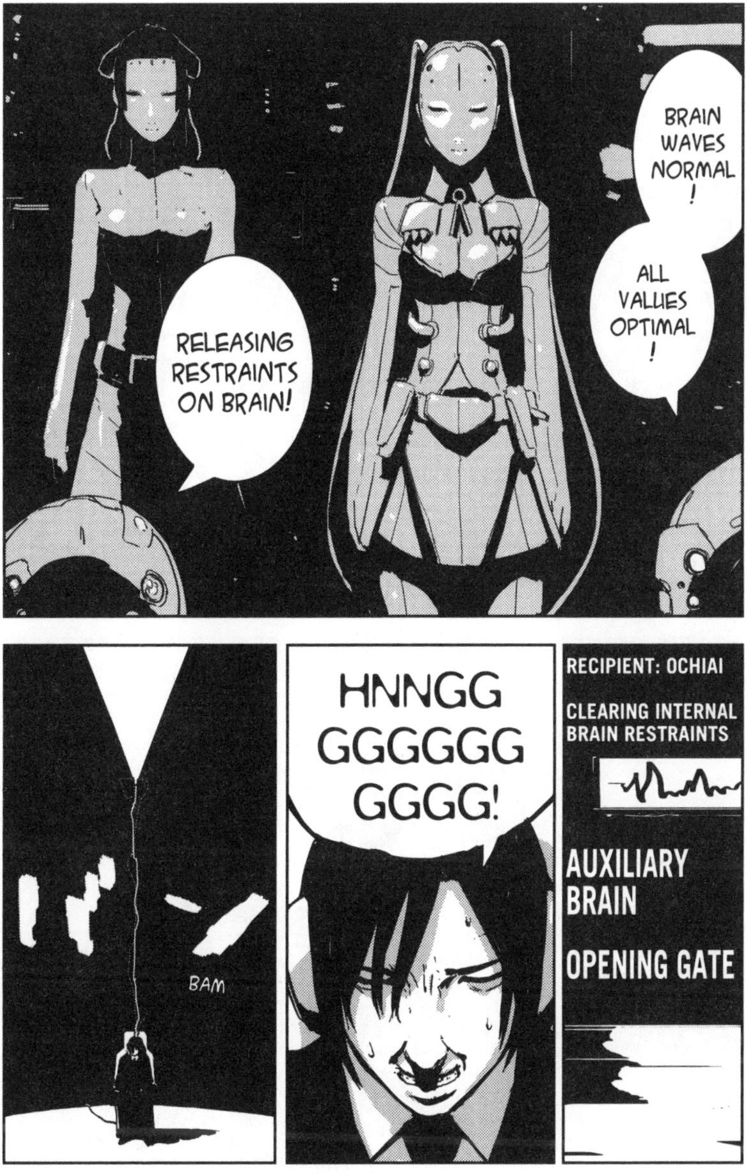
BRAIN WAVES NORMAL!
ALL VALUES OPTIMAL!
RELEASING RESTRAINTS ON BRAIN!
RECIPIENT: OCHIAI
CLEARING INTERNAL BRAIN RESTRAINTS
AUXILIARY BRAIN
OPENING GATE
HNNGG GGGGGG GGGG!
BAM

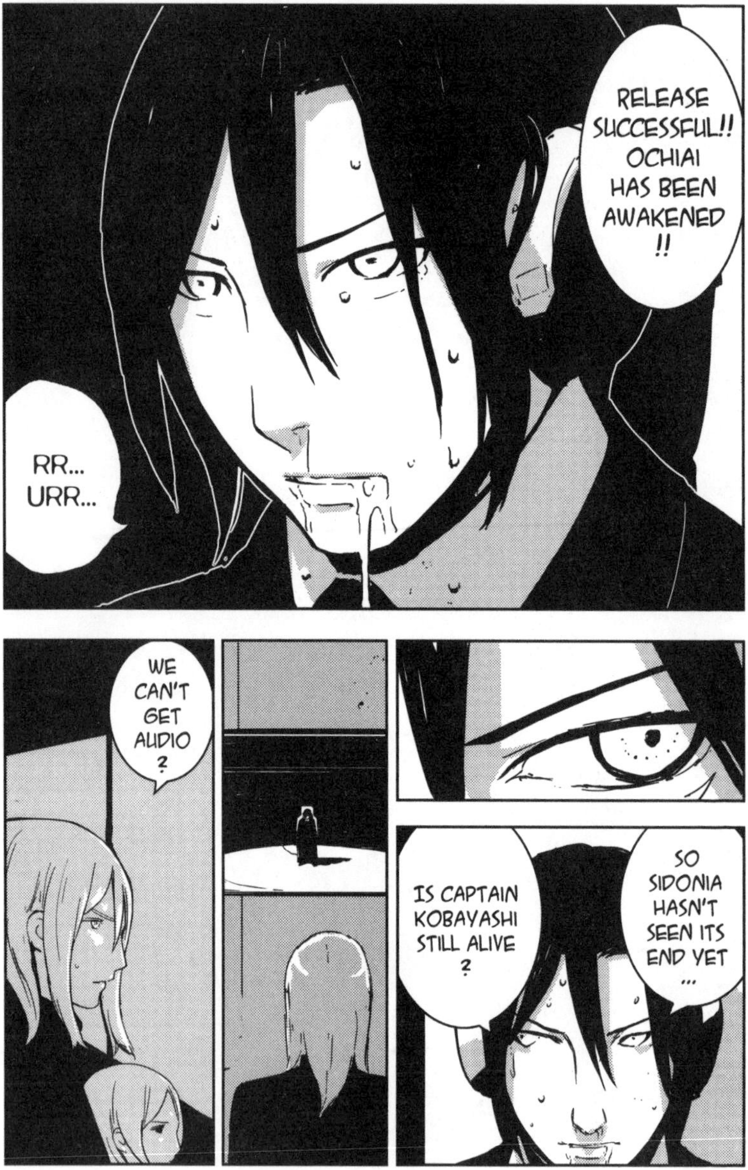
RELEASE SUCCESSFUL!! OCHIAI HAS BEEN AWAKENED !!
RR... URR...
SO SIDONIA HASN'T SEEN ITS END YET ...
IS CAPTAIN KOBAYASHI STILL ALIVE ?
WE CAN'T GET AUDIO ?

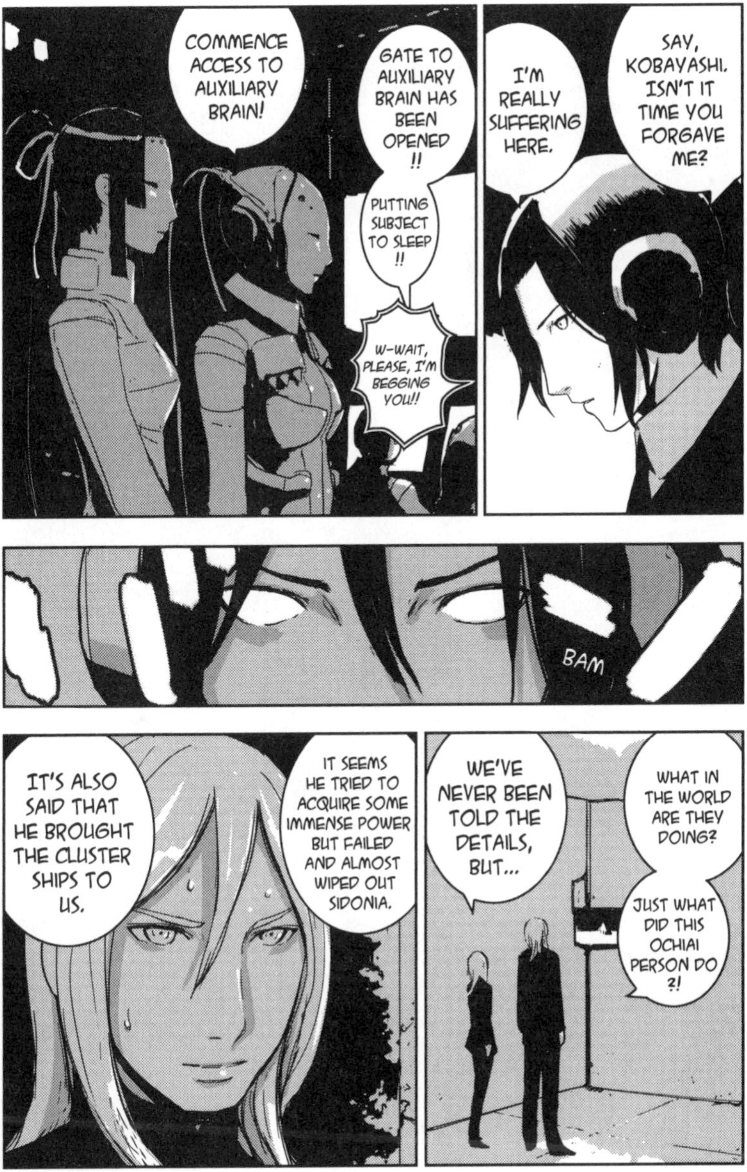
SAY, KOBAYASHI. ISN'T IT TIME YOU FORGAVE ME?
I'M REALLY SUFFERING HERE.
GATE TO AUXILIARY BRAIN HAS BEEN OPENED!!
COMMENCE ACCESS TO AUXILIARY BRAIN!
PUTTING SUBJECT TO SLEEP!!
W-WAIT, PLEASE, I'M BEGGING YOU!!
BAM
WHAT IN THE WORLD ARE THEY DOING?
JUST WHAT DID THIS OCHIAI PERSON DO?!
WE'VE NEVER BEEN TOLD THE DETAILS, BUT...
IT SEEMS HE TRIED TO ACQUIRE SOME IMMENSE POWER BUT FAILED AND ALMOST WIPED OUT SIDONIA.
IT'S ALSO SAID THAT HE BROUGHT THE CLUSTER SHIPS TO US.

Chapter 17: END

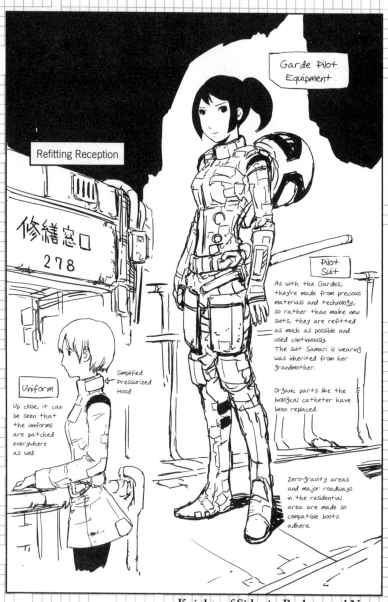

Garde Pilot Equipment

Refitting Reception

修繕窓口
278

Pilot Suit

As with the Gardes, they're made from precious materials and technology, so rather than make new suits, they are refitted as much as possible and used continuously.
The suit Samari is wearing was inherited from her grandmother.

Organic parts like the biological catheter have been replaced.

Zero-gravity areas and major roadways in the residential area are made so compatible boots adhere.

Uniform

Up close, it can be seen that the uniforms are patched everywhere as well.

Simplified Pressurized Hood

Knights of Sidonia Background Notes

Chapter 18: Scientist Ochiai's Daughter

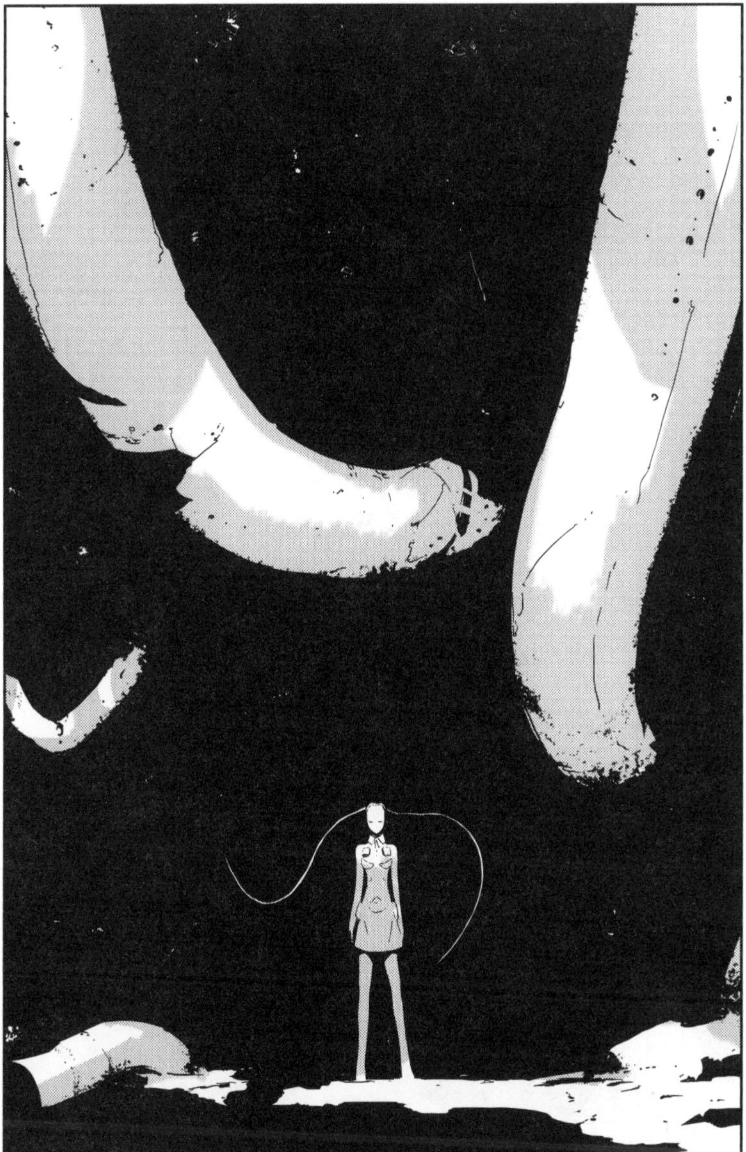

One Hundred Sights of Sidonia Part Fourteen: Underground Plumbing Pathway

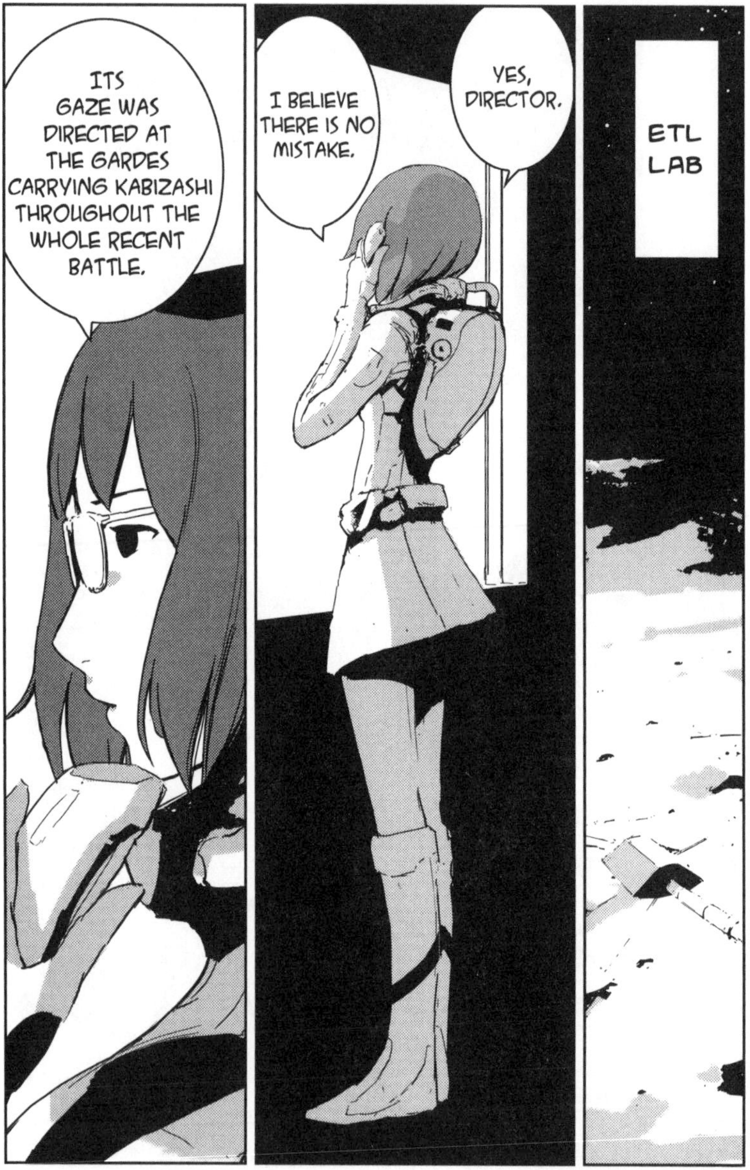
ETL LAB
YES, DIRECTOR.
I BELIEVE THERE IS NO MISTAKE.
ITS GAZE WAS DIRECTED AT THE GARDES CARRYING KABIZASHI THROUGHOUT THE WHOLE RECENT BATTLE.

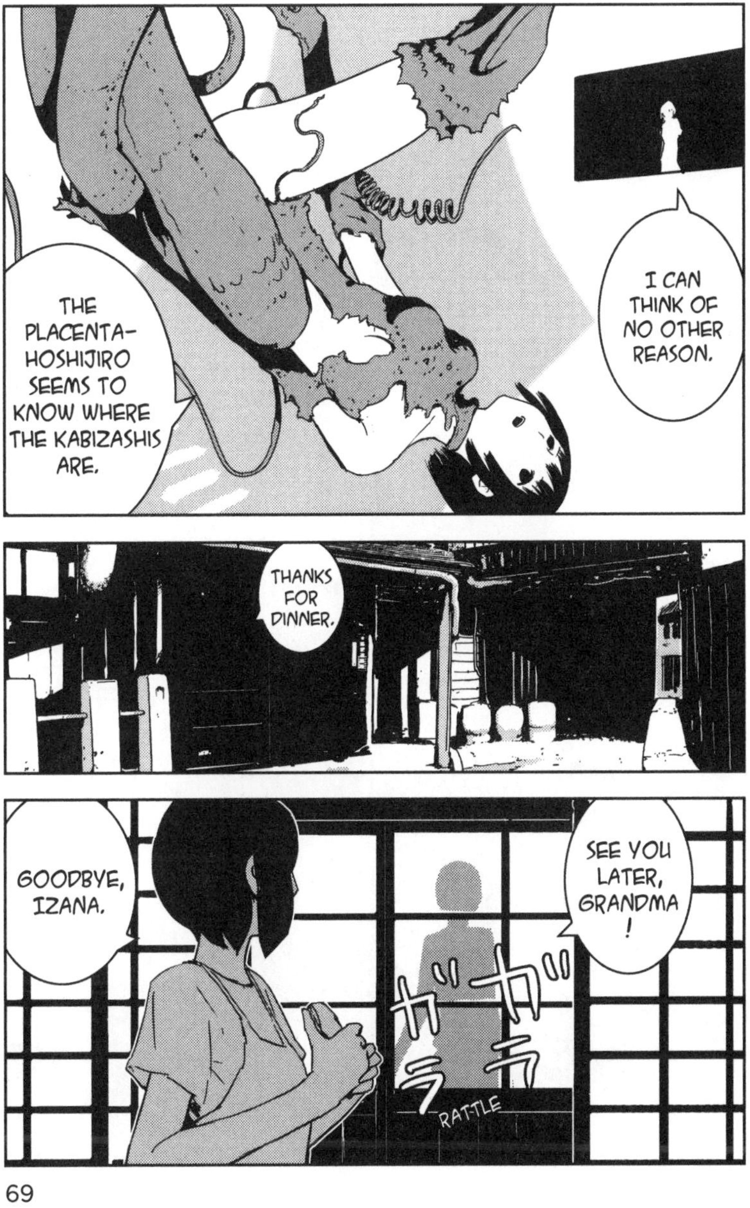
I CAN THINK OF NO OTHER REASON.
THE PLACENTA-HOSHIJIRO SEEMS TO KNOW WHERE THE KABIZASHIS ARE.
THANKS FOR DINNER.
SEE YOU LATER, GRANDMA!
ガラガラ
RATTLE
GOODBYE, IZANA.

YOU'RE THE FIRST, DIRECTOR.

NO.

HANDLE WITH CARE FROM NOW ON TOO.

GOOD... I'M MAKING THIS FIRST-CLASS CLASSIFIED INFO.

Y-YES, UNDER-STOOD.

GCHNK

OOPS!

I WASN'T DONE TUNING UP IZANA'S PROSTHETIC ARM!

BY THE WAY, NUMI, HAVE YOU TOLD ANYONE ELSE?

SORRY I CUT YOU OFF.

70

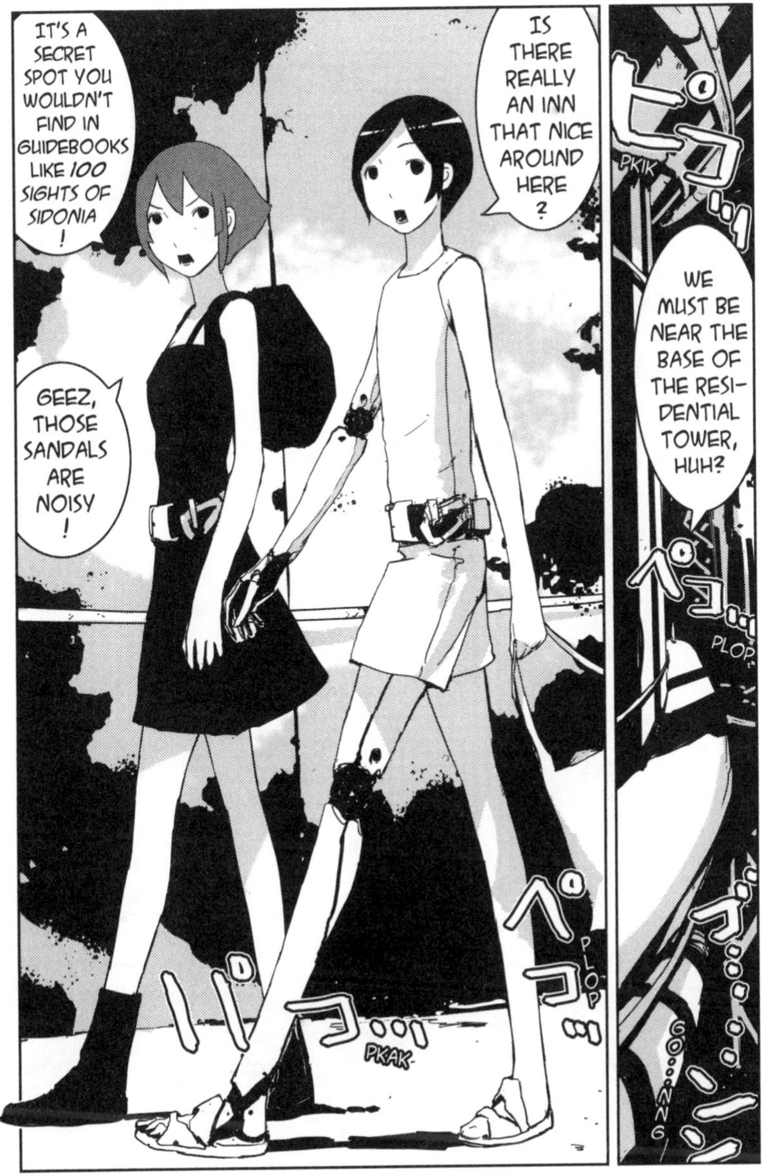
PKIK
WE MUST BE NEAR THE BASE OF THE RESI-DENTIAL TOWER, HUH?
PLOP
GOOONNG
IS THERE REALLY AN INN THAT NICE AROUND HERE?
IT'S A SECRET SPOT YOU WOULDN'T FIND IN GUIDEBOOKS LIKE *100 SIGHTS OF SIDONIA*!
GEEZ, THOSE SANDALS ARE NOISY!
PLOP
PKAK

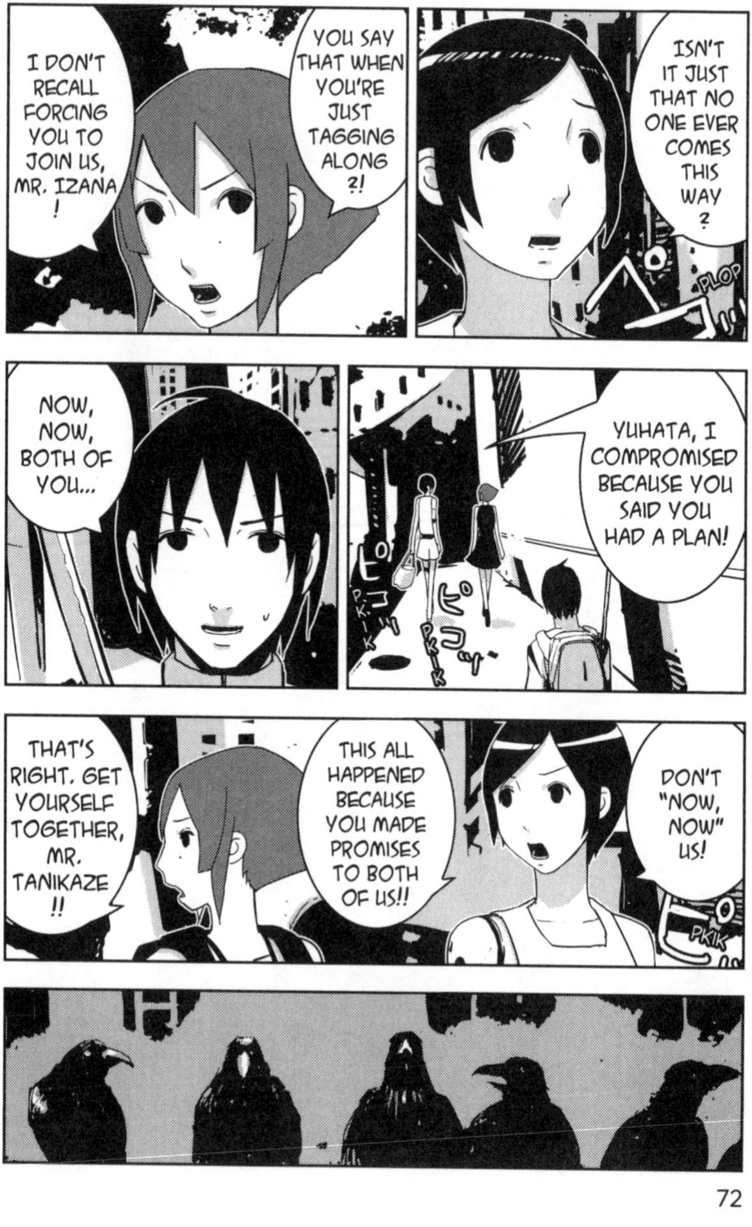
ISN'T IT JUST THAT NO ONE EVER COMES THIS WAY?
PLOP
YOU SAY THAT WHEN YOU'RE JUST TAGGING ALONG?!
I DON'T RECALL FORCING YOU TO JOIN US, MR. IZANA!
YUHATA, I COMPROMISED BECAUSE YOU SAID YOU HAD A PLAN!
PKIK
PKIK
NOW, NOW, BOTH OF YOU...
DON'T "NOW, NOW" US!
PKIK
THIS ALL HAPPENED BECAUSE YOU MADE PROMISES TO BOTH OF US!!
THAT'S RIGHT. GET YOURSELF TOGETHER, MR. TANIKAZE!!

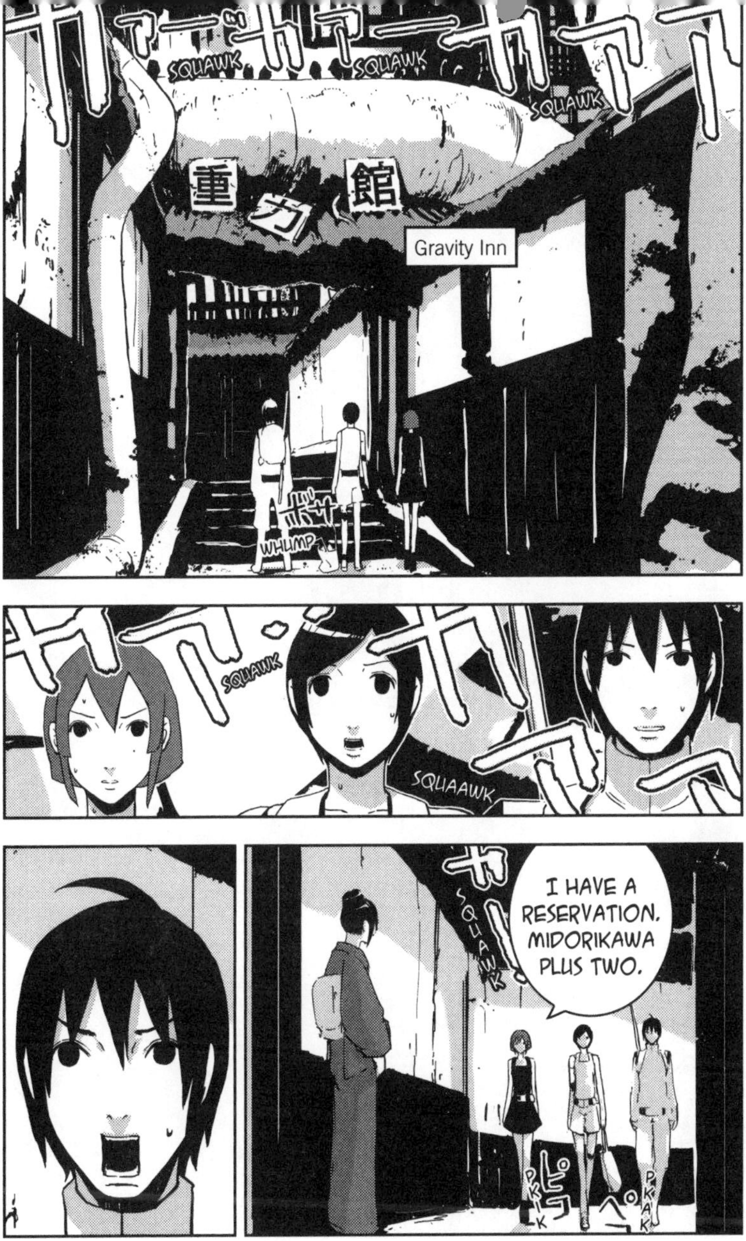
SQUAWK
SQUAWK
SQUAWK
重力館
Gravity Inn
ボサ
WHUMP
SQUAWK
SQUAAWK
SQUAWK
I HAVE A RESERVATION. MIDORIKAWA PLUS TWO.
PKAK
PKIK

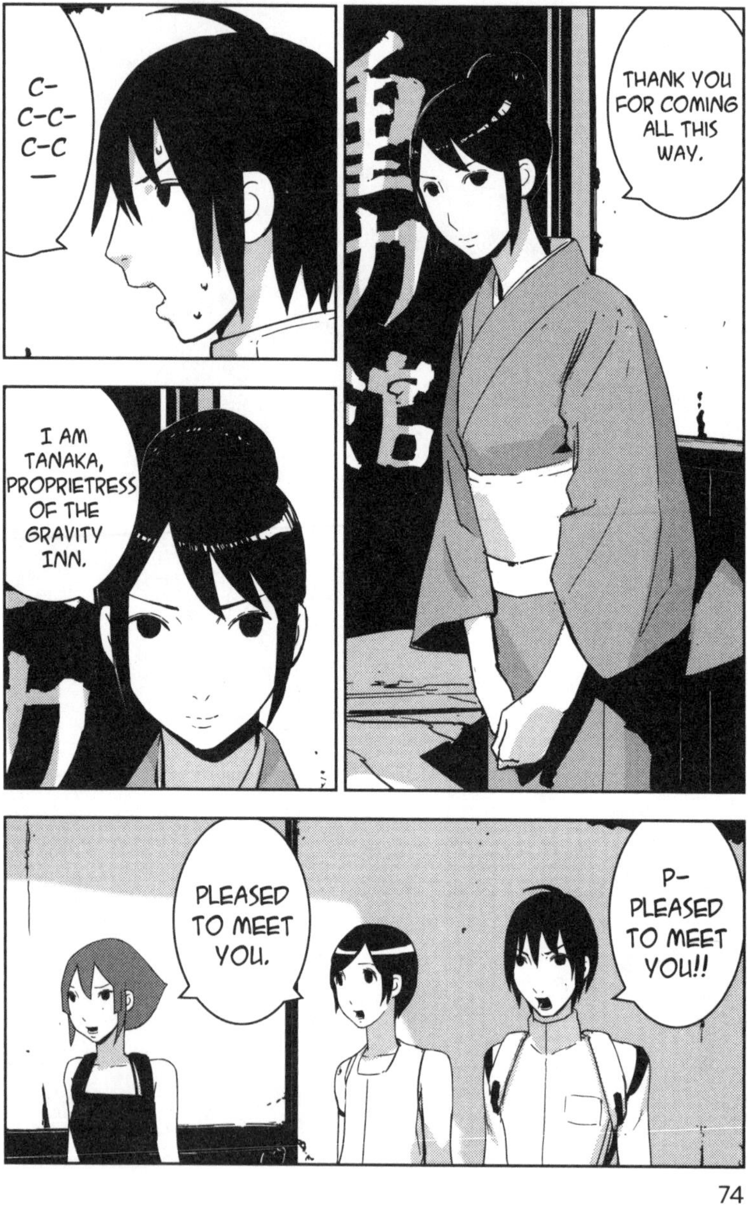
THANK YOU FOR COMING ALL THIS WAY.
C- C-C- C-C —
I AM TANAKA, PROPRIETRESS OF THE GRAVITY INN.
P- PLEASED TO MEET YOU!!
PLEASED TO MEET YOU.

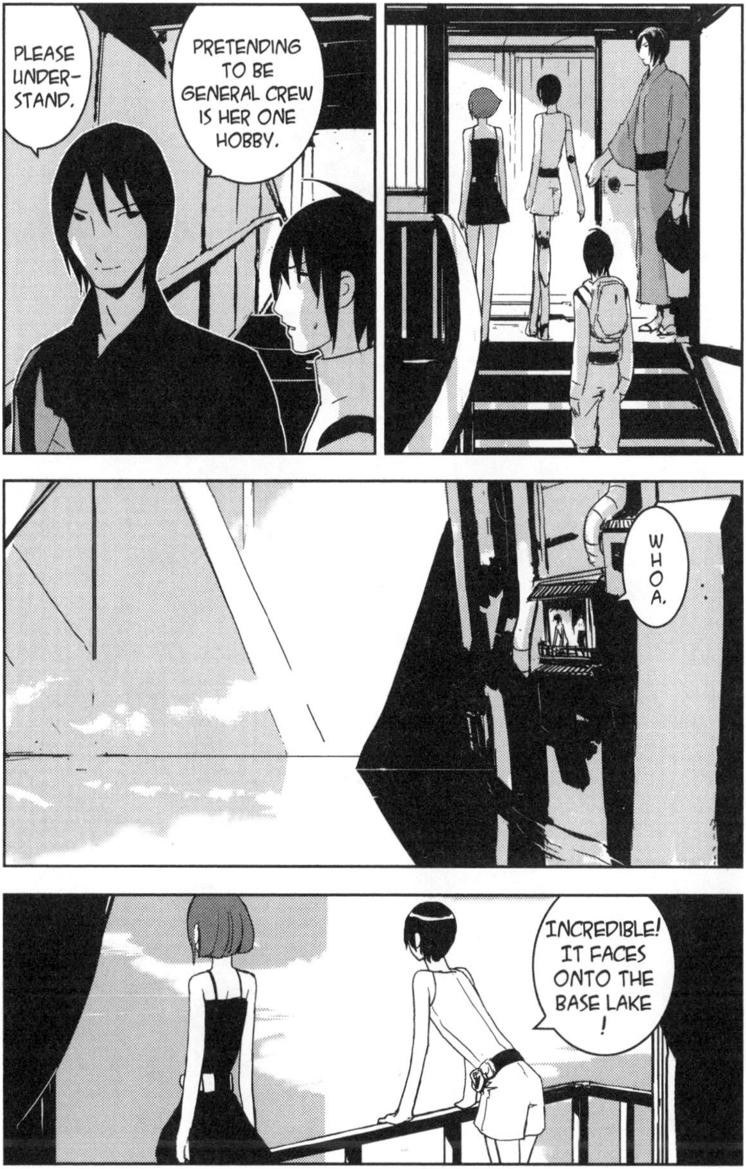
PRETENDING TO BE GENERAL CREW IS HER ONE HOBBY.
PLEASE UNDER-STAND.
WHOA.
INCREDIBLE! IT FACES ONTO THE BASE LAKE!

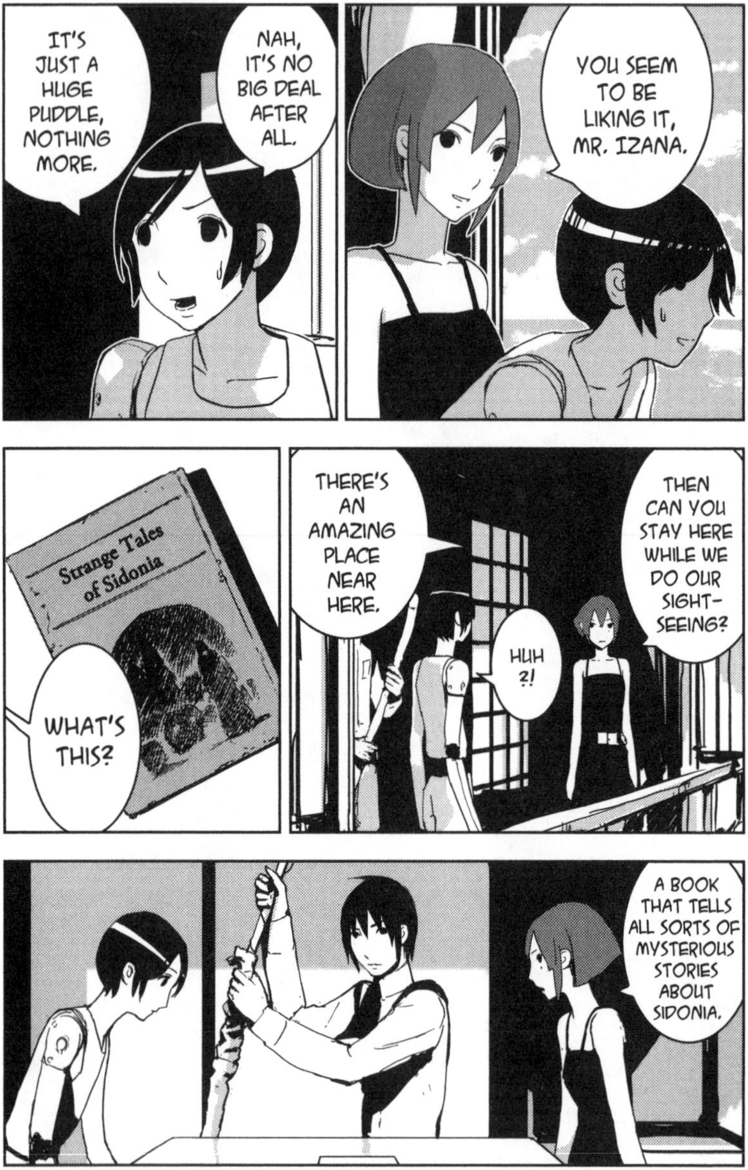
YOU SEEM TO BE LIKING IT, MR. IZANA.
NAH, IT'S NO BIG DEAL AFTER ALL.
IT'S JUST A HUGE PUDDLE, NOTHING MORE.
THEN CAN YOU STAY HERE WHILE WE DO OUR SIGHT-SEEING?
THERE'S AN AMAZING PLACE NEAR HERE.
HUH ?!
Strange Tales of Sidonia
WHAT'S THIS?
A BOOK THAT TELLS ALL SORTS OF MYSTERIOUS STORIES ABOUT SIDONIA.

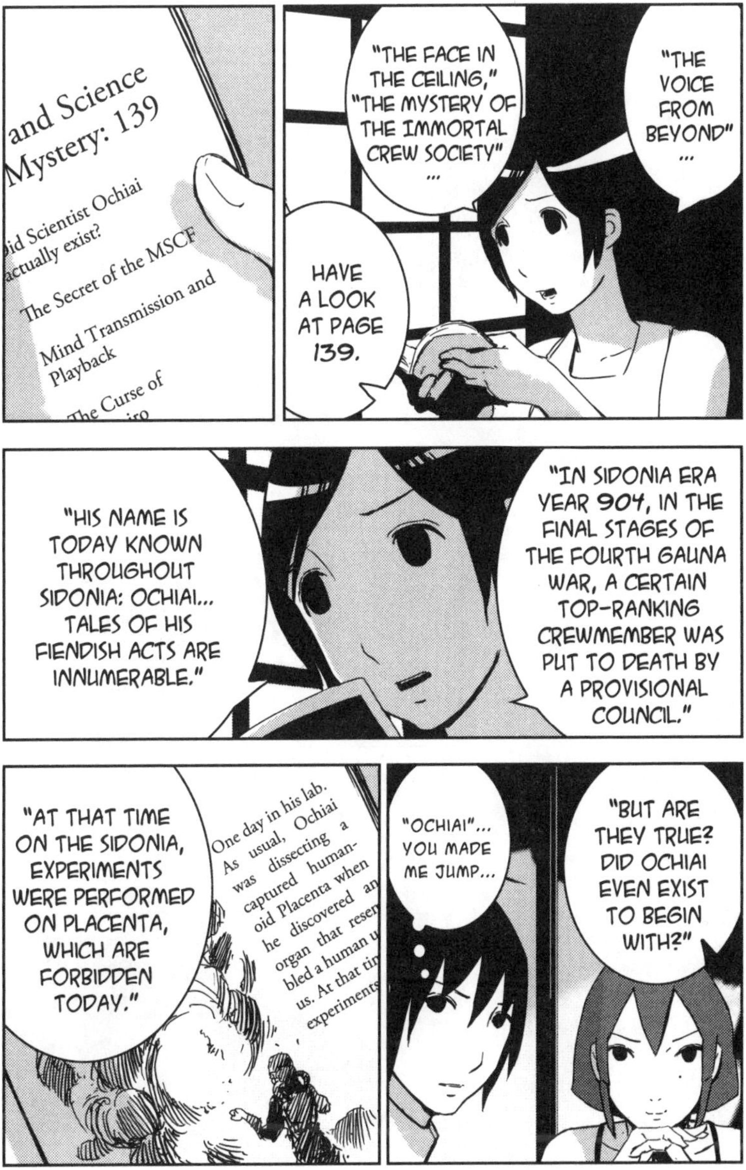
"THE VOICE FROM BEYOND"...
"THE FACE IN THE CEILING," "THE MYSTERY OF THE IMMORTAL CREW SOCIETY"...
HAVE A LOOK AT PAGE 139.
and Science
Mystery: 139
id Scientist Ochiai
actually exist?
The Secret of the MSCF
Mind Transmission and
Playback
The Curse of
"IN SIDONIA ERA YEAR 904, IN THE FINAL STAGES OF THE FOURTH GAUNA WAR, A CERTAIN TOP-RANKING CREWMEMBER WAS PUT TO DEATH BY A PROVISIONAL COUNCIL."
"HIS NAME IS TODAY KNOWN THROUGHOUT SIDONIA: OCHIAI... TALES OF HIS FIENDISH ACTS ARE INNUMERABLE."
"BUT ARE THEY TRUE? DID OCHIAI EVEN EXIST TO BEGIN WITH?"
"OCHIAI"... YOU MADE ME JUMP...
One day in his lab.
As usual, Ochiai
was dissecting a
captured human-
oid Placenta when
he discovered an
organ that resem
bled a human u
us. At that tim
experiments
"AT THAT TIME ON THE SIDONIA, EXPERIMENTS WERE PERFORMED ON PLACENTA, WHICH ARE FORBIDDEN TODAY."

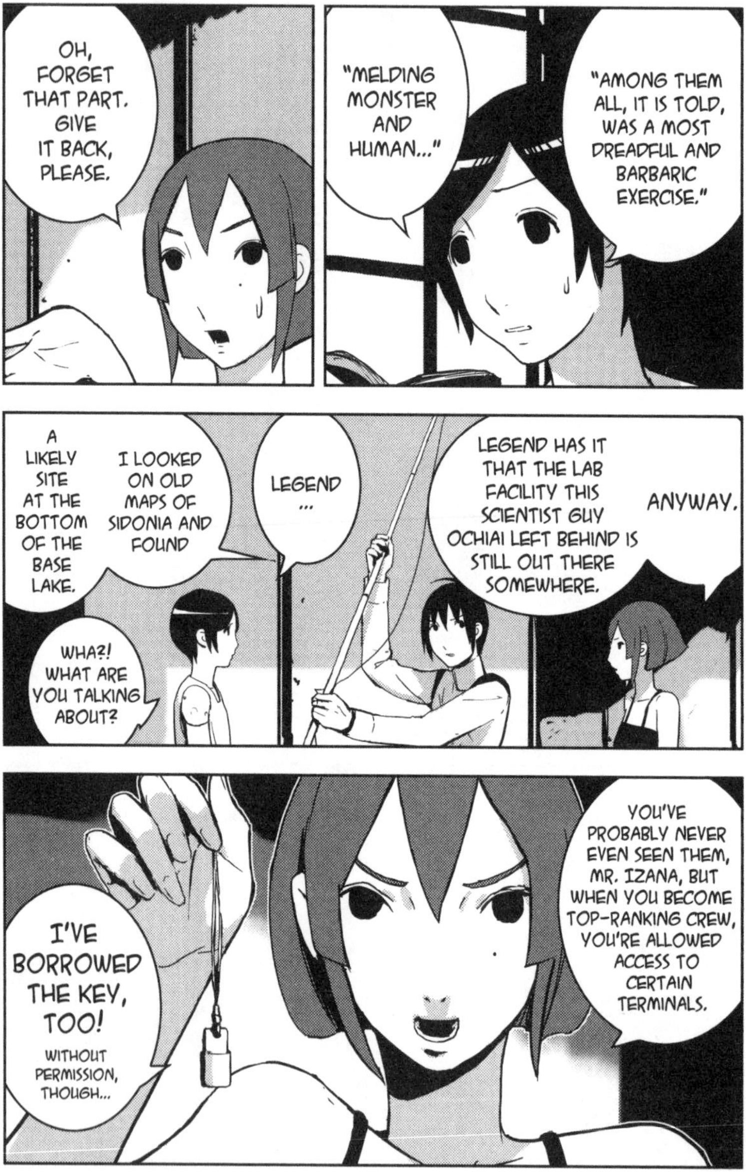
"AMONG THEM ALL, IT IS TOLD, WAS A MOST DREADFUL AND BARBARIC EXERCISE."
"MELDING MONSTER AND HUMAN..."
OH, FORGET THAT PART. GIVE IT BACK, PLEASE.
ANYWAY.
LEGEND HAS IT THAT THE LAB FACILITY THIS SCIENTIST GUY OCHIAI LEFT BEHIND IS STILL OUT THERE SOMEWHERE.
LEGEND...
I LOOKED ON OLD MAPS OF SIDONIA AND FOUND
A LIKELY SITE AT THE BOTTOM OF THE BASE LAKE.
WHA?! WHAT ARE YOU TALKING ABOUT?
YOU'VE PROBABLY NEVER EVEN SEEN THEM, MR. IZANA, BUT WHEN YOU BECOME TOP-RANKING CREW, YOU'RE ALLOWED ACCESS TO CERTAIN TERMINALS.
I'VE BORROWED THE KEY, TOO!
WITHOUT PERMISSION, THOUGH...

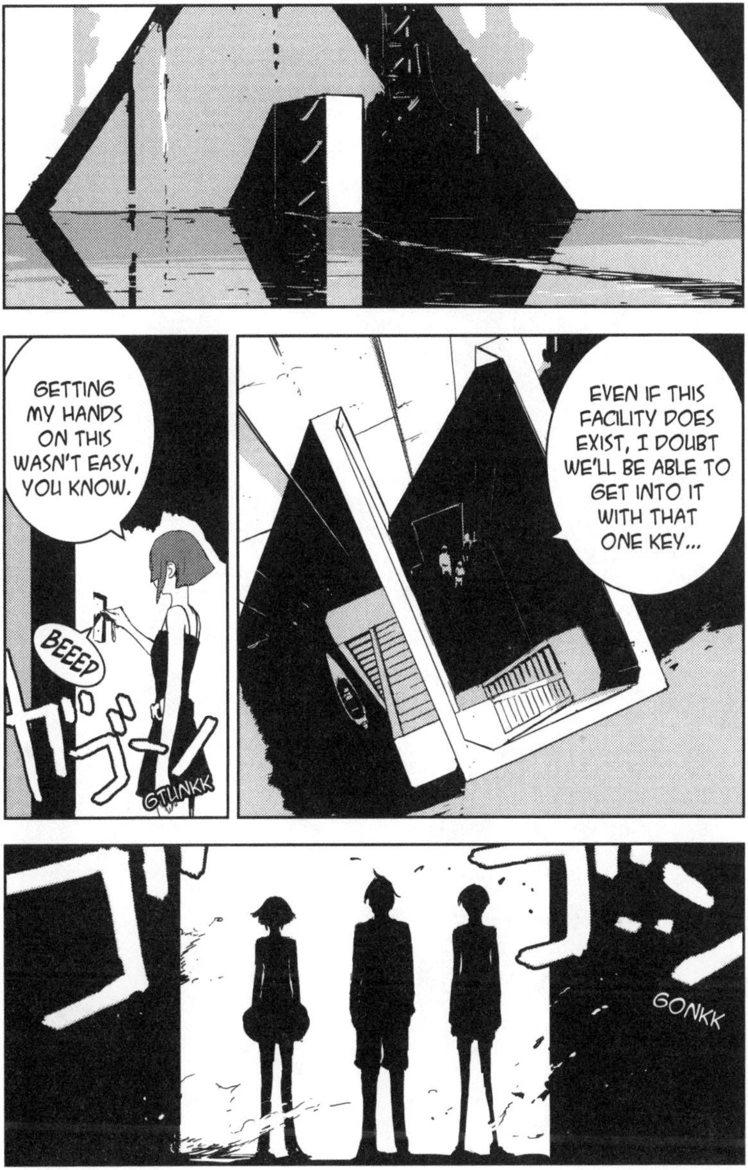
EVEN IF THIS FACILITY DOES EXIST, I DOUBT WE'LL BE ABLE TO GET INTO IT WITH THAT ONE KEY...
GETTING MY HANDS ON THIS WASN'T EASY, YOU KNOW.
BEEEP
GTUNKK
GONKK

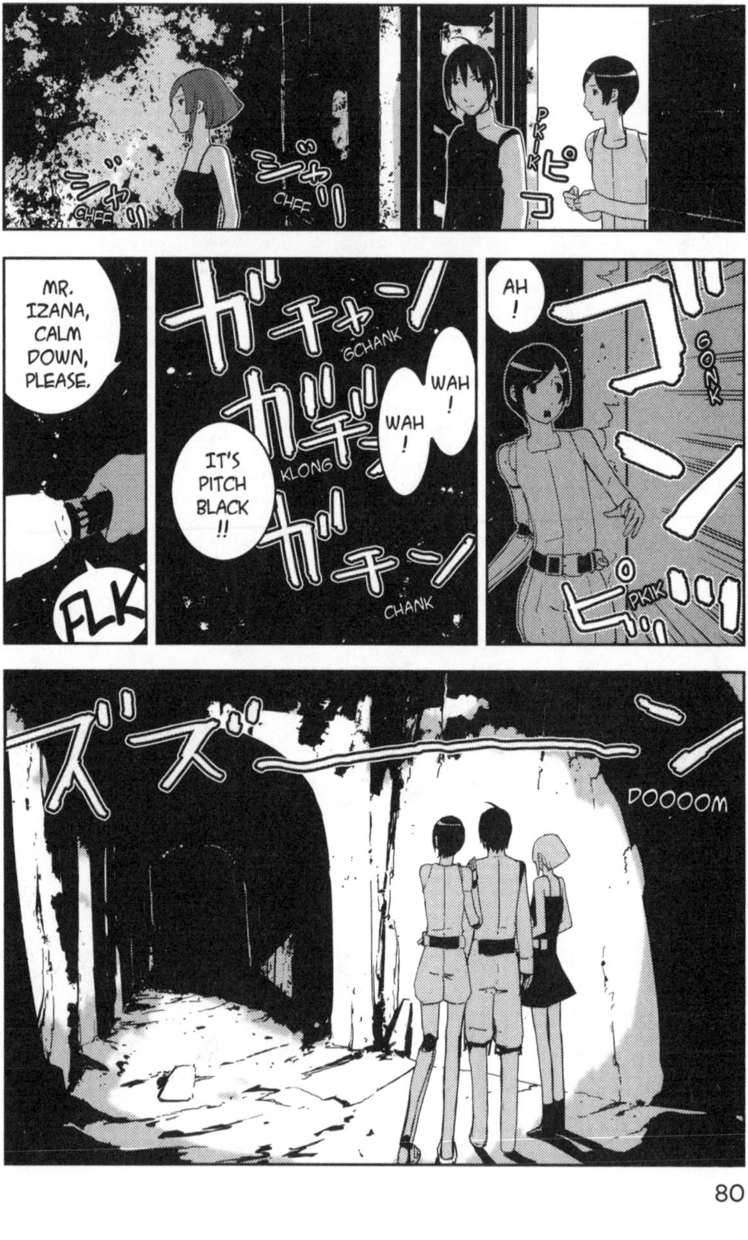
PKIK
CHFF
CHFF
AH!
GONK
PKIK
GCHANK
WAH!
WAH!
KLONG
IT'S PITCH BLACK!!
CHANK
MR. IZANA, CALM DOWN, PLEASE.
FLK
DOOOOM

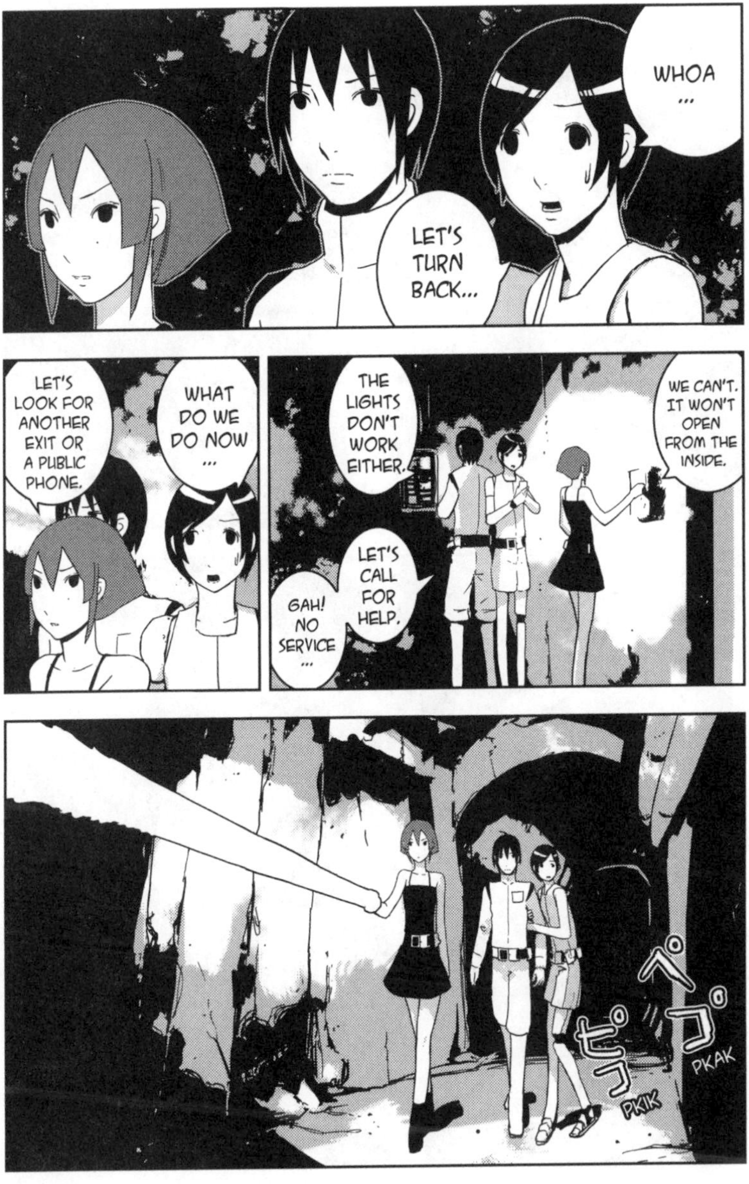
WHOA ...
LET'S TURN BACK...
WE CAN'T. IT WON'T OPEN FROM THE INSIDE.
THE LIGHTS DON'T WORK EITHER.
LET'S CALL FOR HELP.
GAH! NO SERVICE ...
WHAT DO WE DO NOW ...
LET'S LOOK FOR ANOTHER EXIT OR A PUBLIC PHONE.
PKAK
PKIK

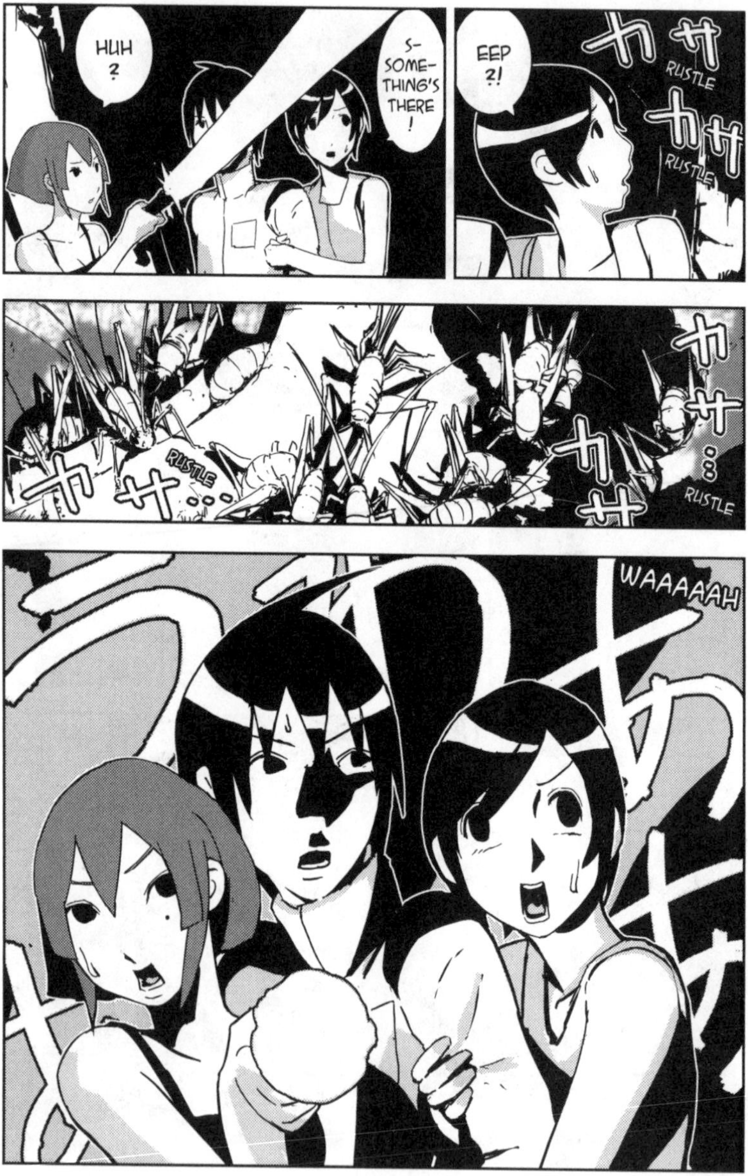
EEP ?!
RUSTLE
RUSTLE
S-SOMETHING'S THERE !
HUH ?
RUSTLE
RUSTLE
WAAAAAH

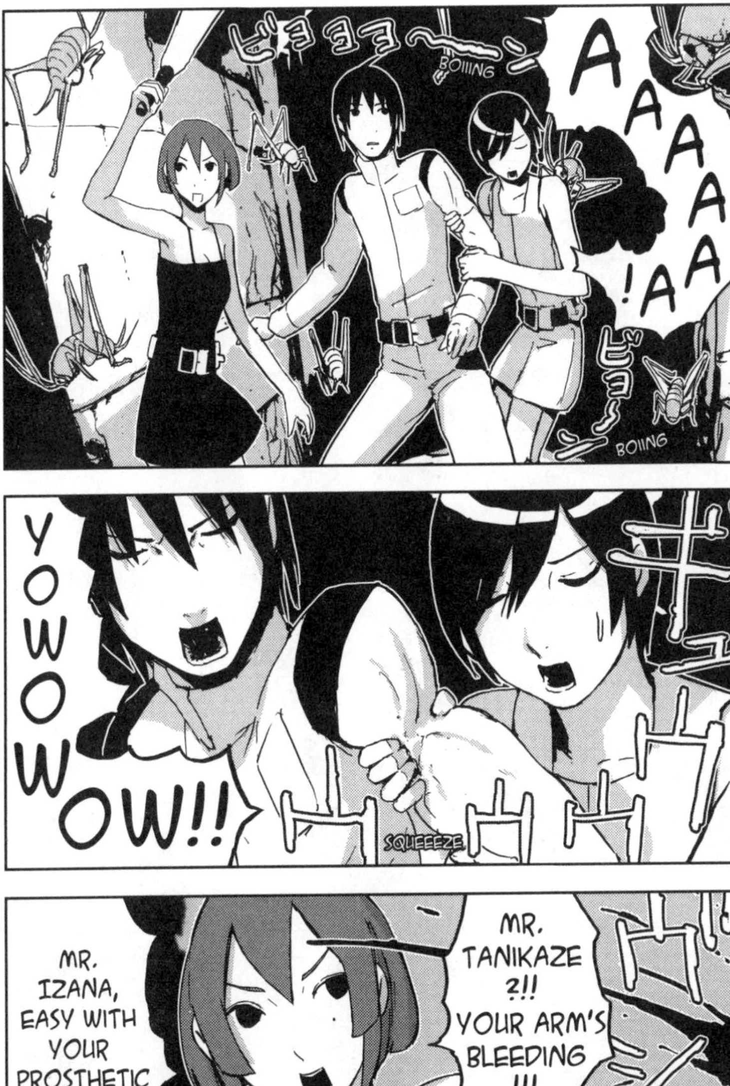
AAAAAA!
BOIING
BOIIING
SQUEEEZE
YOWOWOW!!
MR. TANIKAZE?!! YOUR ARM'S BLEEDING!!!
MR. IZANA, EASY WITH YOUR PROSTHETIC ARM!!
SPLUTCH

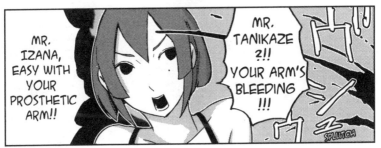

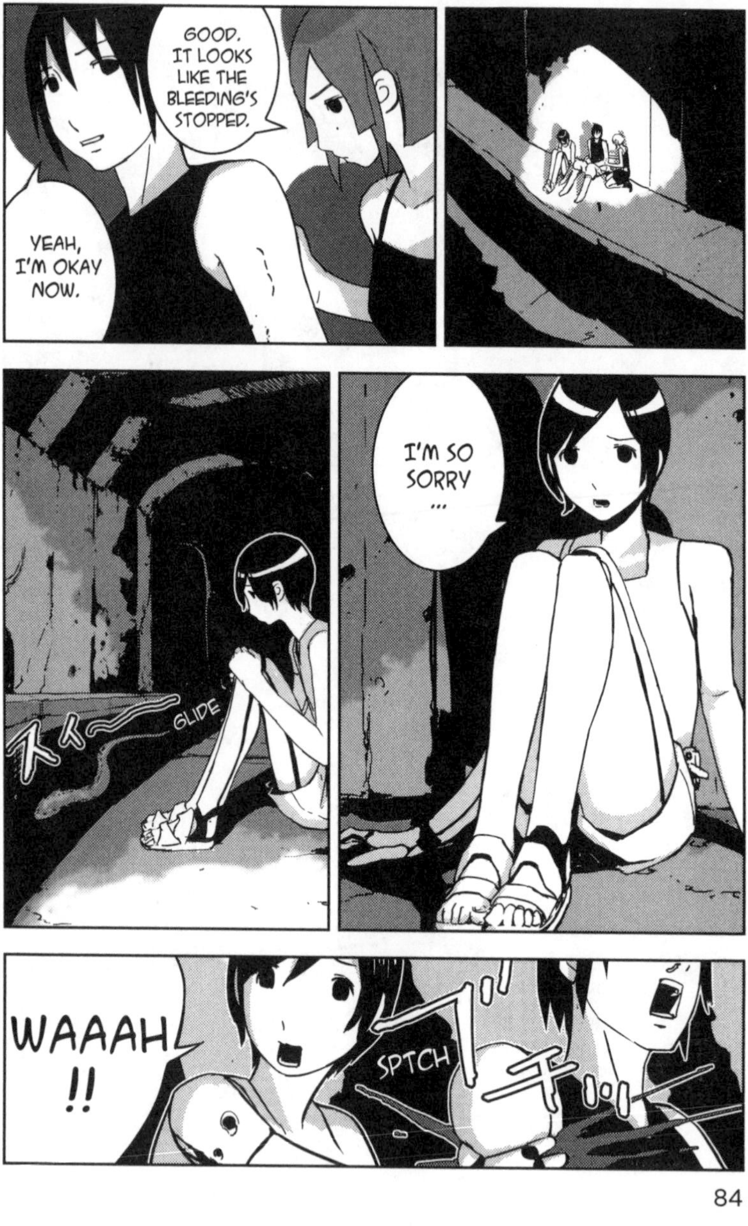
GOOD. IT LOOKS LIKE THE BLEEDING'S STOPPED.
YEAH, I'M OKAY NOW.
I'M SO SORRY ...
スイ〜〜
GLIDE
SPTCH
WAAAH !!

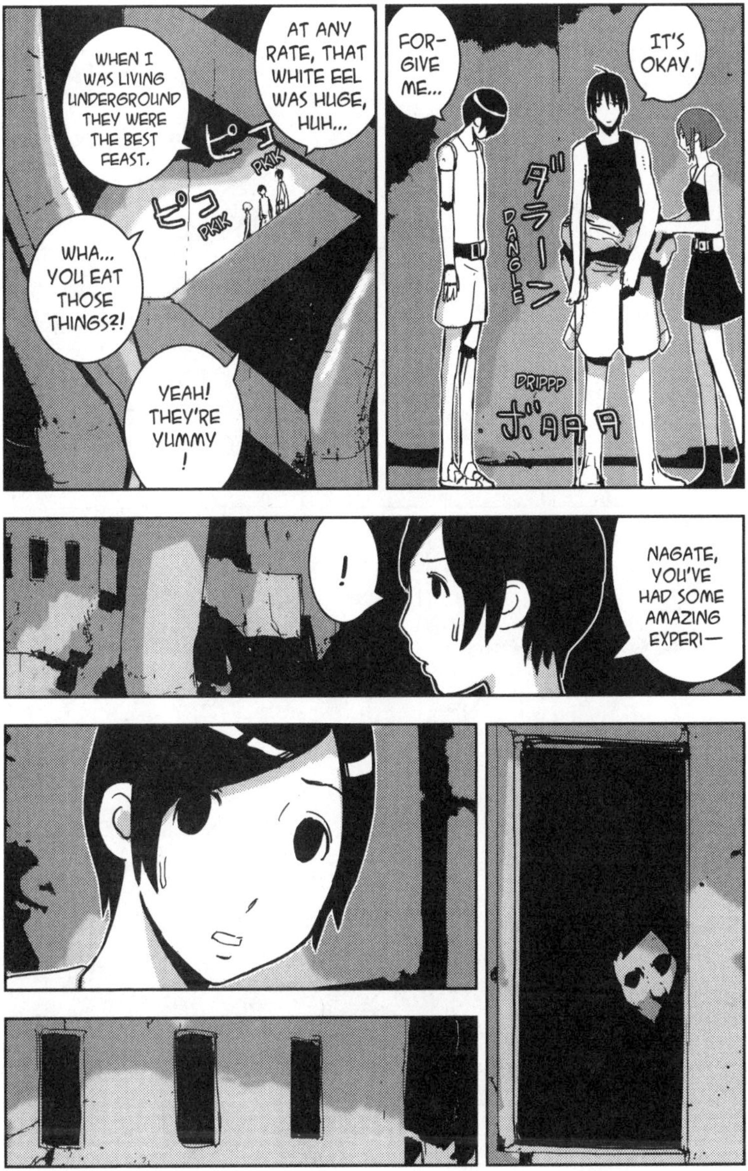
IT'S OKAY.
FOR-GIVE ME...
ダラーン
DANGLE
DRIPPP
ボタタタ
AT ANY RATE, THAT WHITE EEL WAS HUGE, HUH...
WHEN I WAS LIVING UNDERGROUND THEY WERE THE BEST FEAST.
ピコ
PKIK
ピコ
PKIK
WHA... YOU EAT THOSE THINGS?!
YEAH! THEY'RE YUMMY!
NAGATE, YOU'VE HAD SOME AMAZING EXPERI—
!

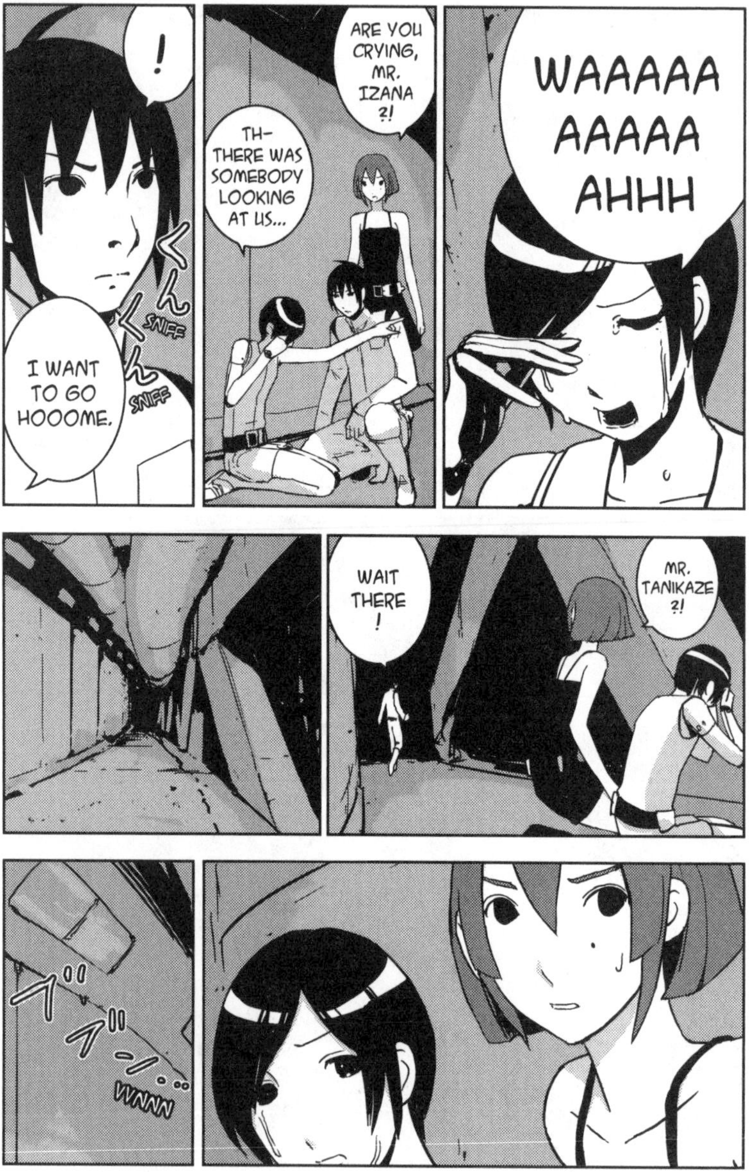
WAAAAA AAAAA AHHH
ARE YOU CRYING, MR. IZANA ?!
TH- THERE WAS SOMEBODY LOOKING AT US...
!
くん SNIFF
くん SNIFF
I WANT TO GO HOOOME.
MR. TANIKAZE ?!
WAIT THERE !
WNNN

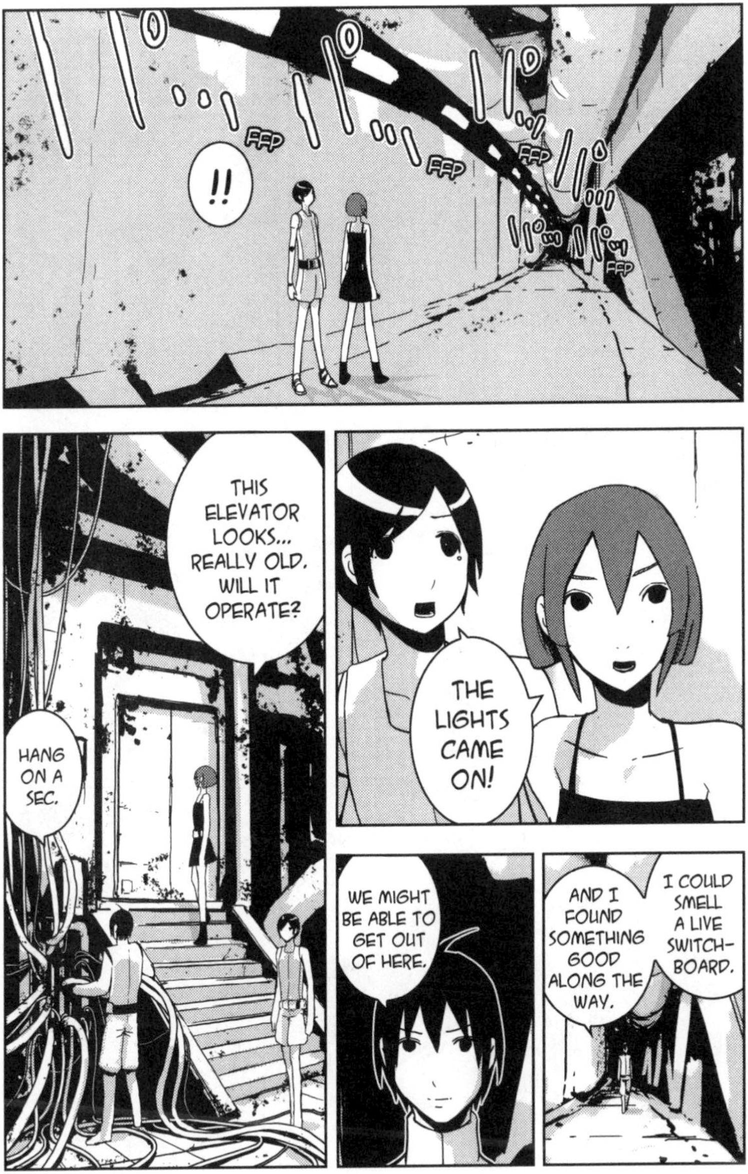
FFP
FFP
FFP
FFP
!!
THE LIGHTS CAME ON!
I COULD SMELL A LIVE SWITCH-BOARD.
AND I FOUND SOMETHING GOOD ALONG THE WAY.
WE MIGHT BE ABLE TO GET OUT OF HERE.
THIS ELEVATOR LOOKS... REALLY OLD. WILL IT OPERATE?
HANG ON A SEC.

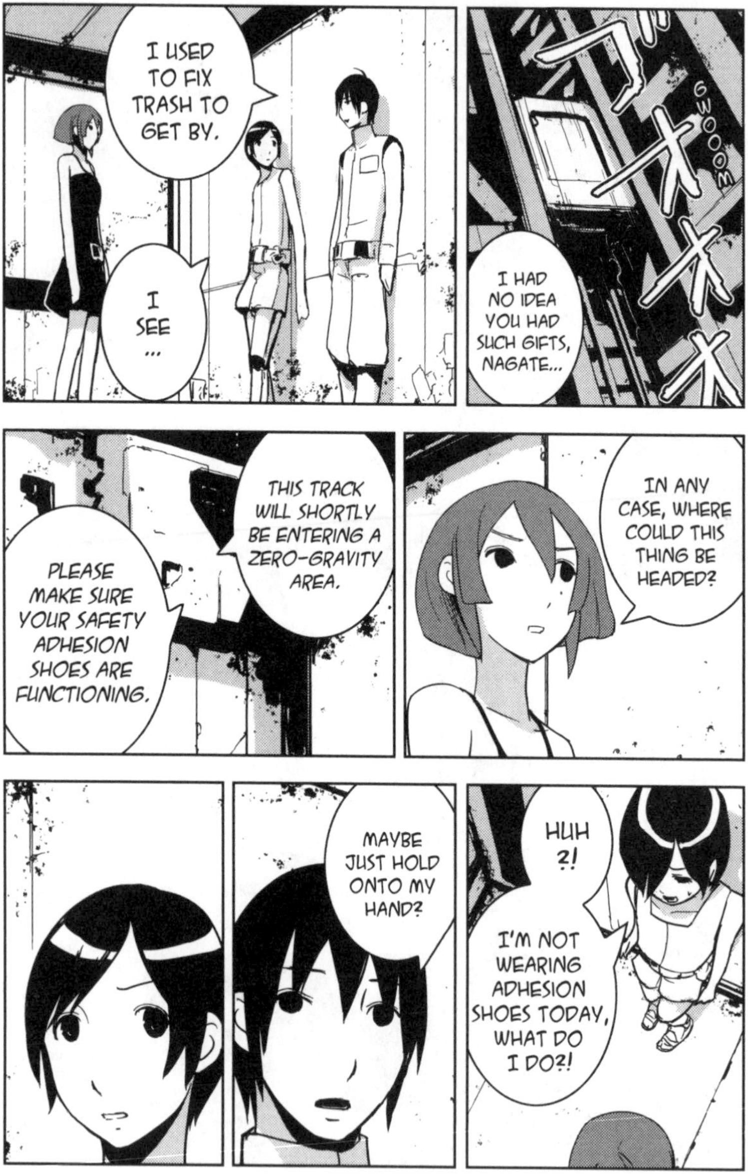
GWOOOOM
I HAD NO IDEA YOU HAD SUCH GIFTS, NAGATE...
I USED TO FIX TRASH TO GET BY.
I SEE ...
IN ANY CASE, WHERE COULD THIS THING BE HEADED?
THIS TRACK WILL SHORTLY BE ENTERING A ZERO-GRAVITY AREA.
PLEASE MAKE SURE YOUR SAFETY ADHESION SHOES ARE FUNCTIONING.
HUH ?!
I'M NOT WEARING ADHESION SHOES TODAY, WHAT DO I DO?!
MAYBE JUST HOLD ONTO MY HAND?

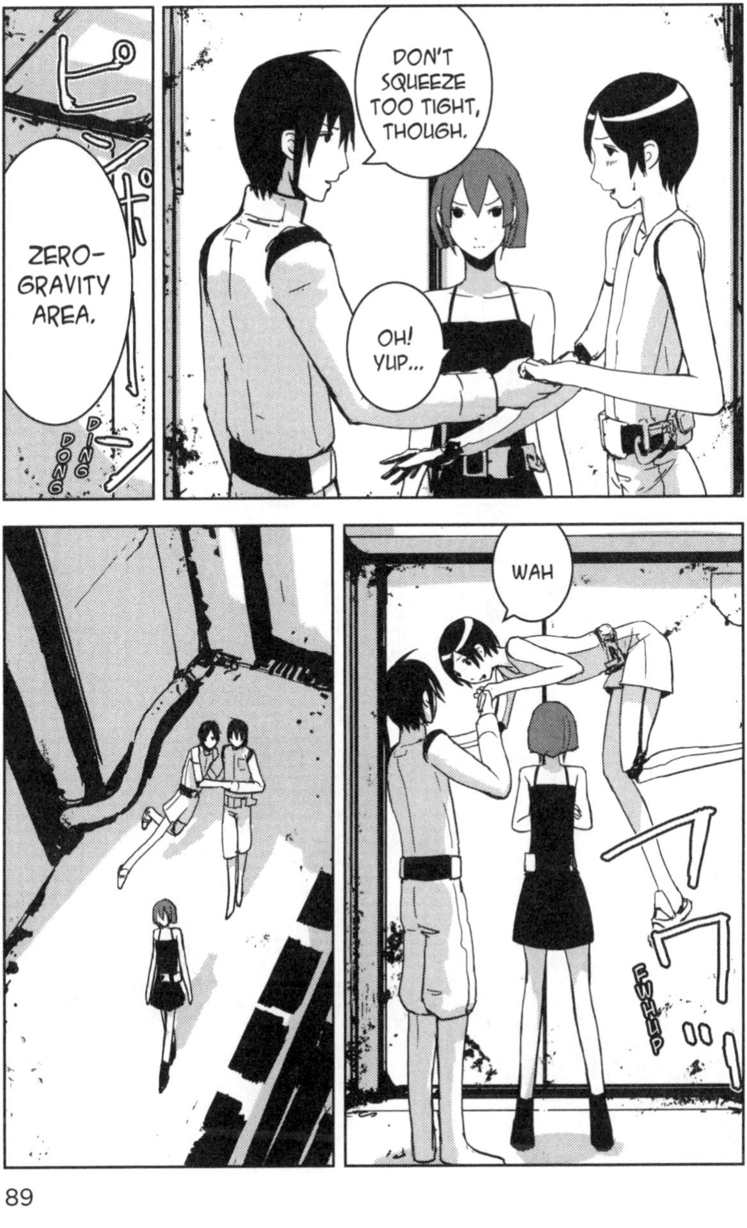
DON'T SQUEEZE TOO TIGHT, THOUGH.
OH! YUP...
ZERO-GRAVITY AREA.
DING DONG
WAH
FWHUP

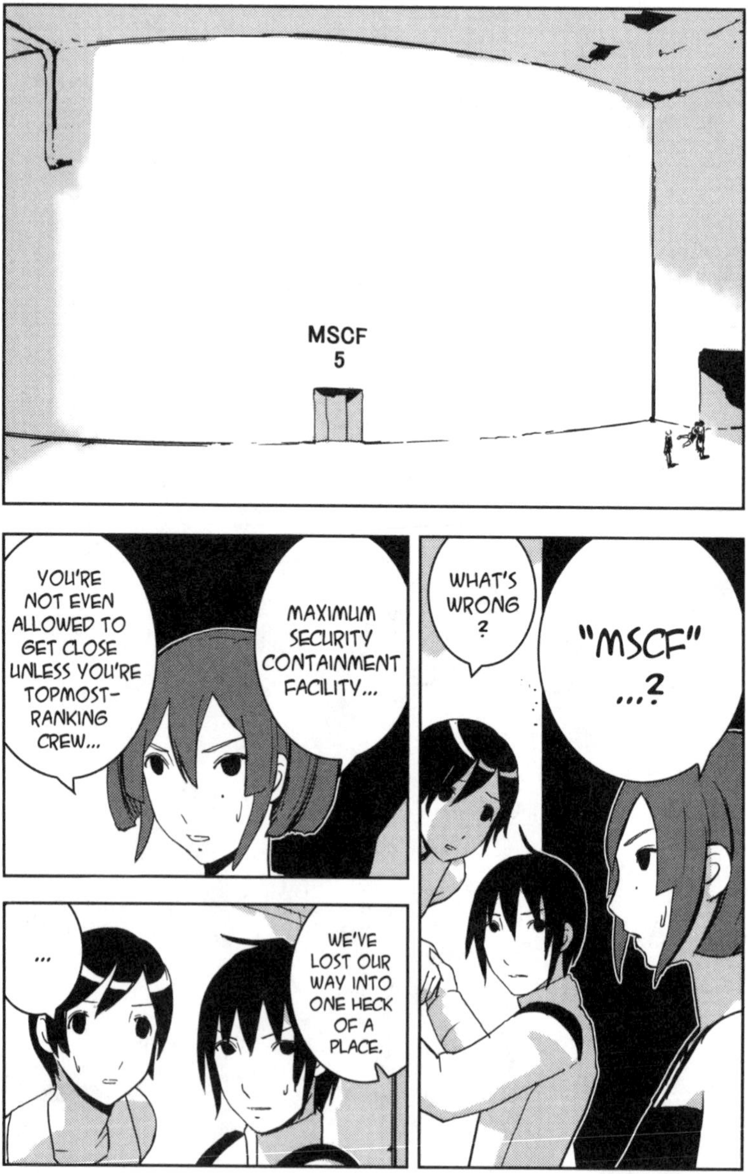
MSCF
5
"MSCF" ...?
WHAT'S WRONG?
MAXIMUM SECURITY CONTAINMENT FACILITY...
YOU'RE NOT EVEN ALLOWED TO GET CLOSE UNLESS YOU'RE TOPMOST-RANKING CREW...
WE'VE LOST OUR WAY INTO ONE HECK OF A PLACE.
...

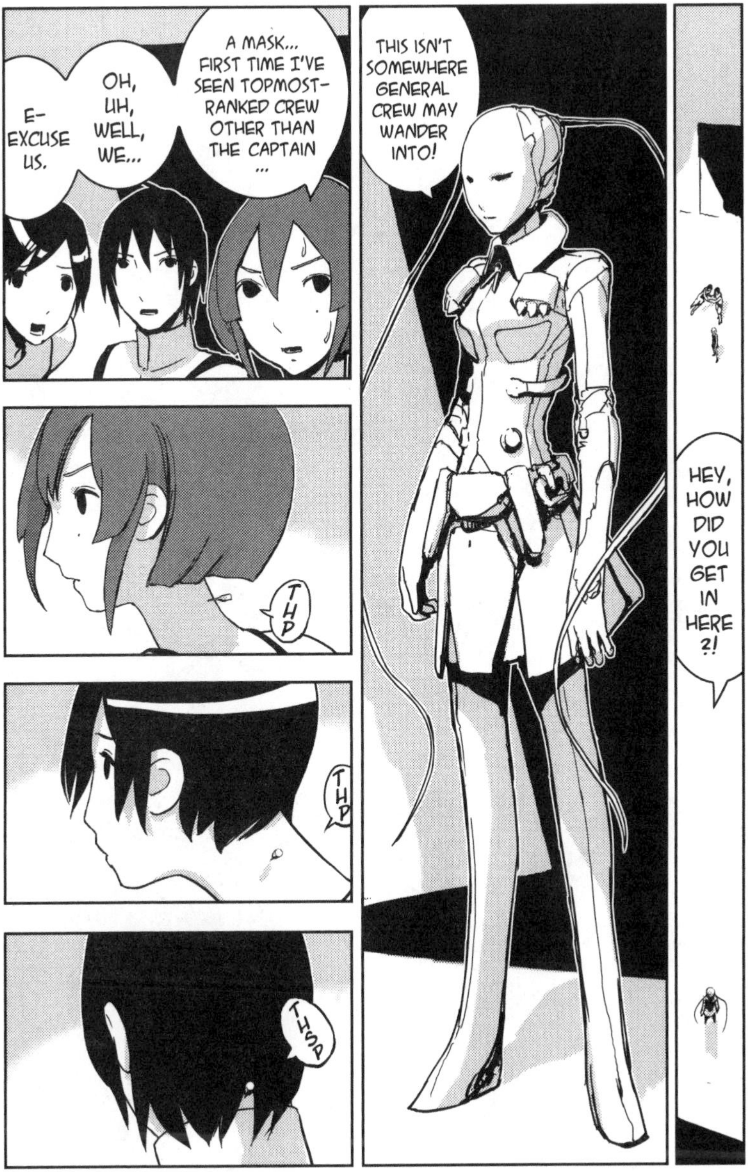
HEY, HOW DID YOU GET IN HERE?!
THIS ISN'T SOMEWHERE GENERAL CREW MAY WANDER INTO!
A MASK... FIRST TIME I'VE SEEN TOPMOST-RANKED CREW OTHER THAN THE CAPTAIN...
OH, UH, WELL, WE...
E-EXCUSE US.
THP
THP
THSP

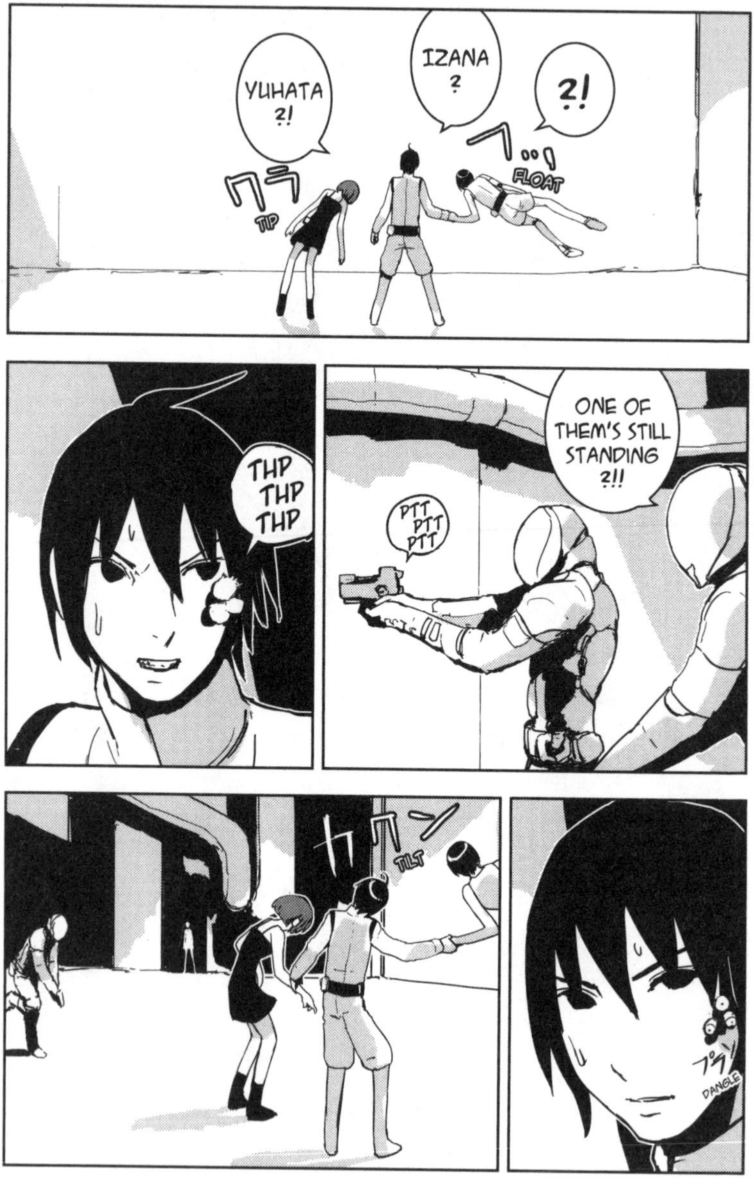
?!
FLOAT
IZANA ?
YUHATA ?!
TIP
ONE OF THEM'S STILL STANDING ?!!
PTT PTT PTT
THP THP THP
DANGLE
TILT

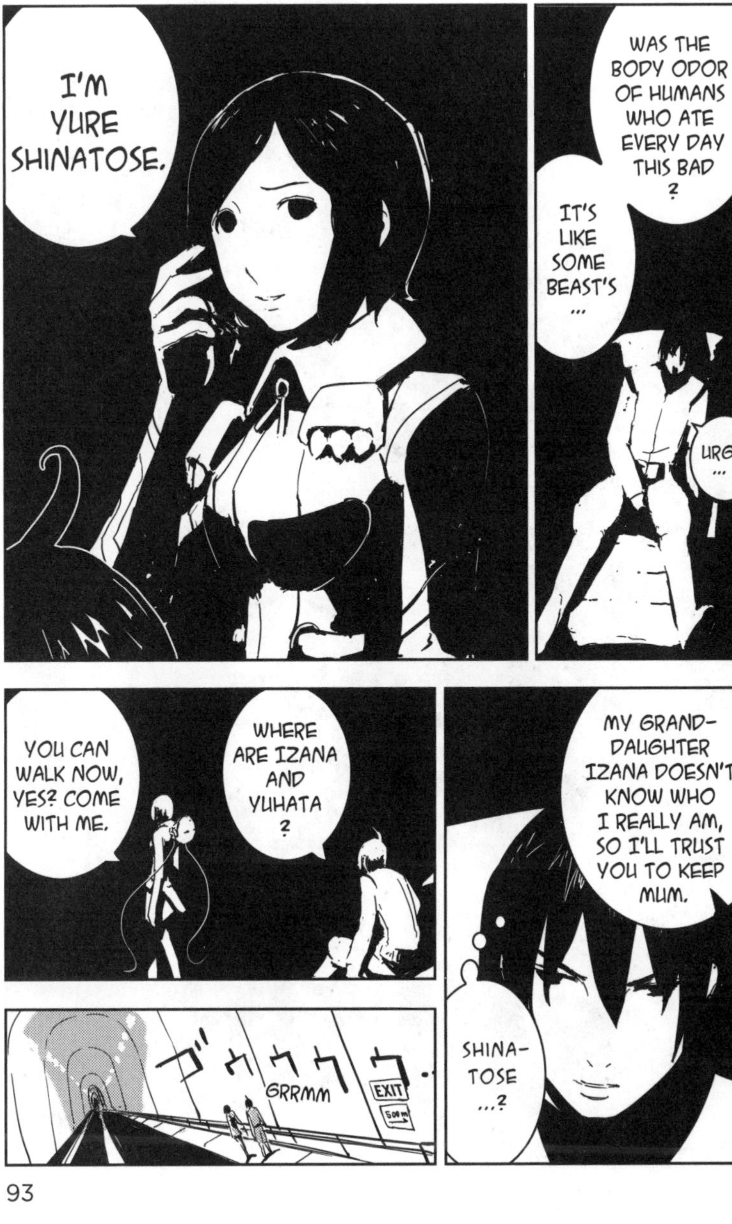

WAS THE BODY ODOR OF HUMANS WHO ATE EVERY DAY THIS BAD?
IT'S LIKE SOME BEAST'S...
URG...
I'M YURE SHINATOSE.
MY GRAND-DAUGHTER IZANA DOESN'T KNOW WHO I REALLY AM, SO I'LL TRUST YOU TO KEEP MUM.
SHINA-TOSE...?
WHERE ARE IZANA AND YUHATA?
YOU CAN WALK NOW, YES? COME WITH ME.
GRRMM
EXIT
500m

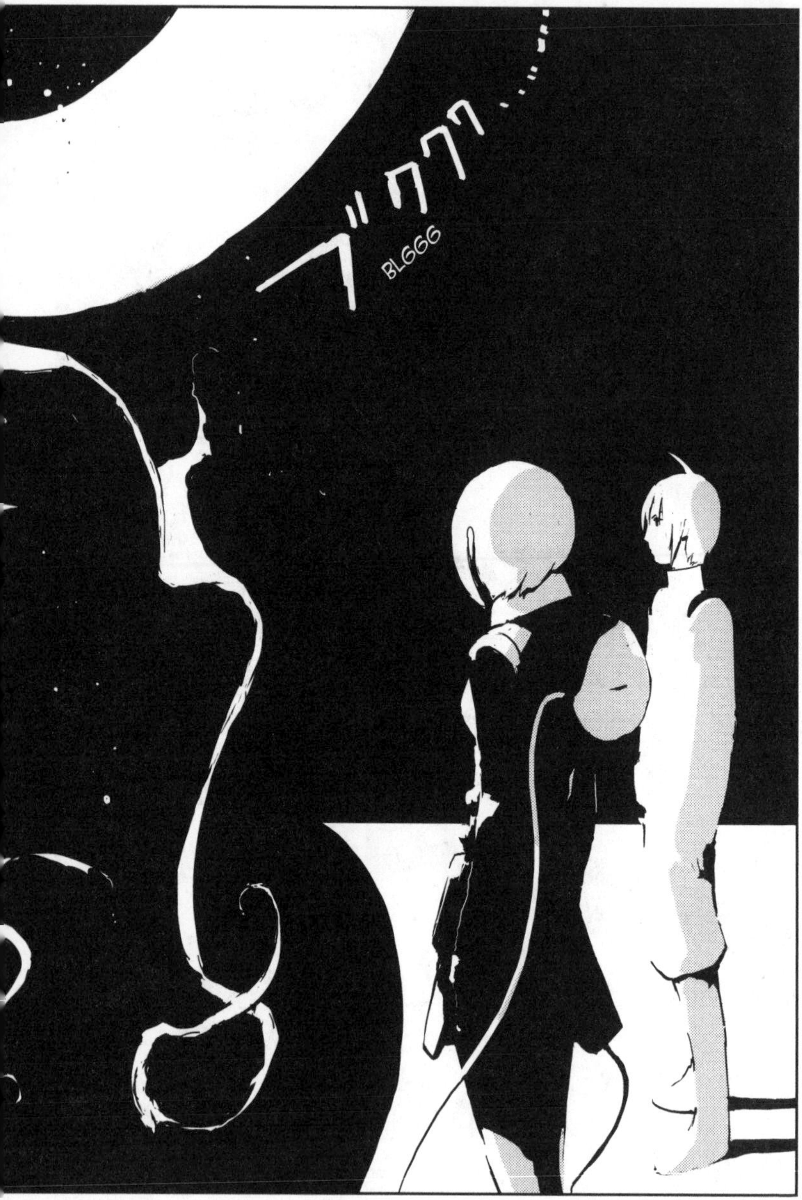
ブゥゥ……
BLGGG

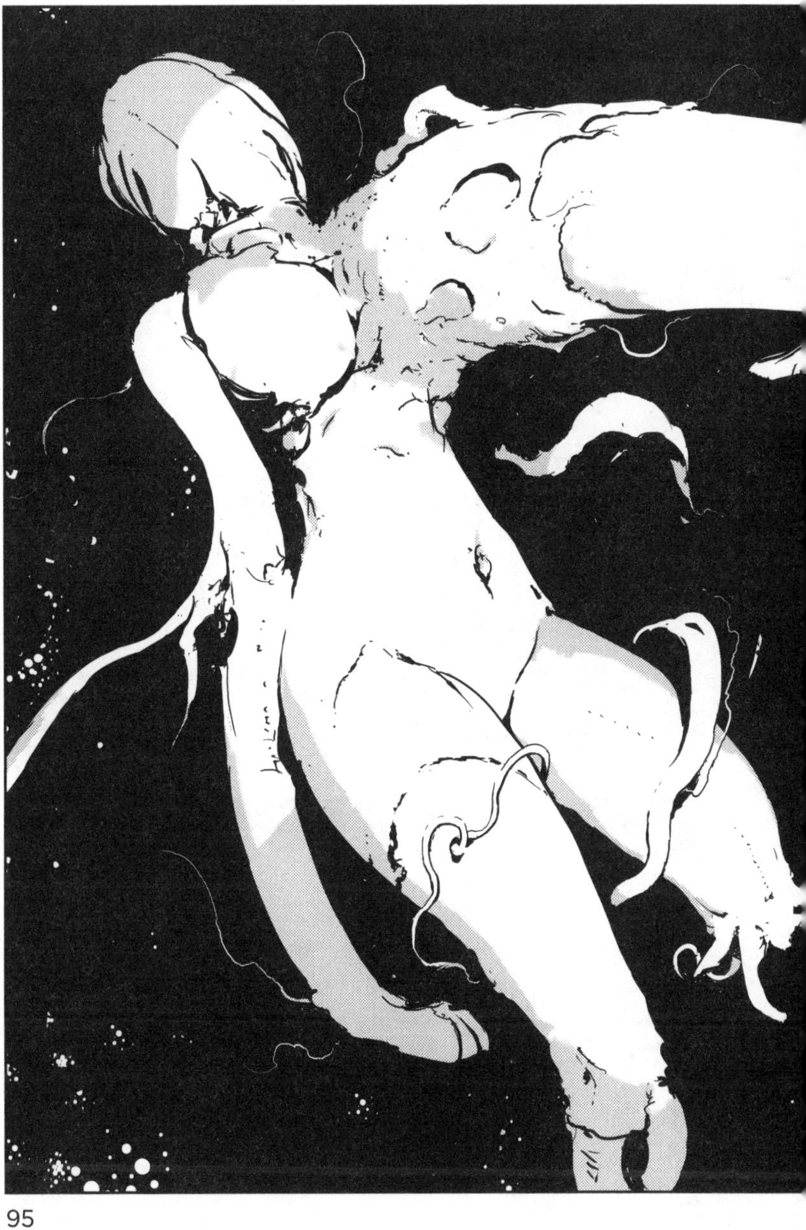

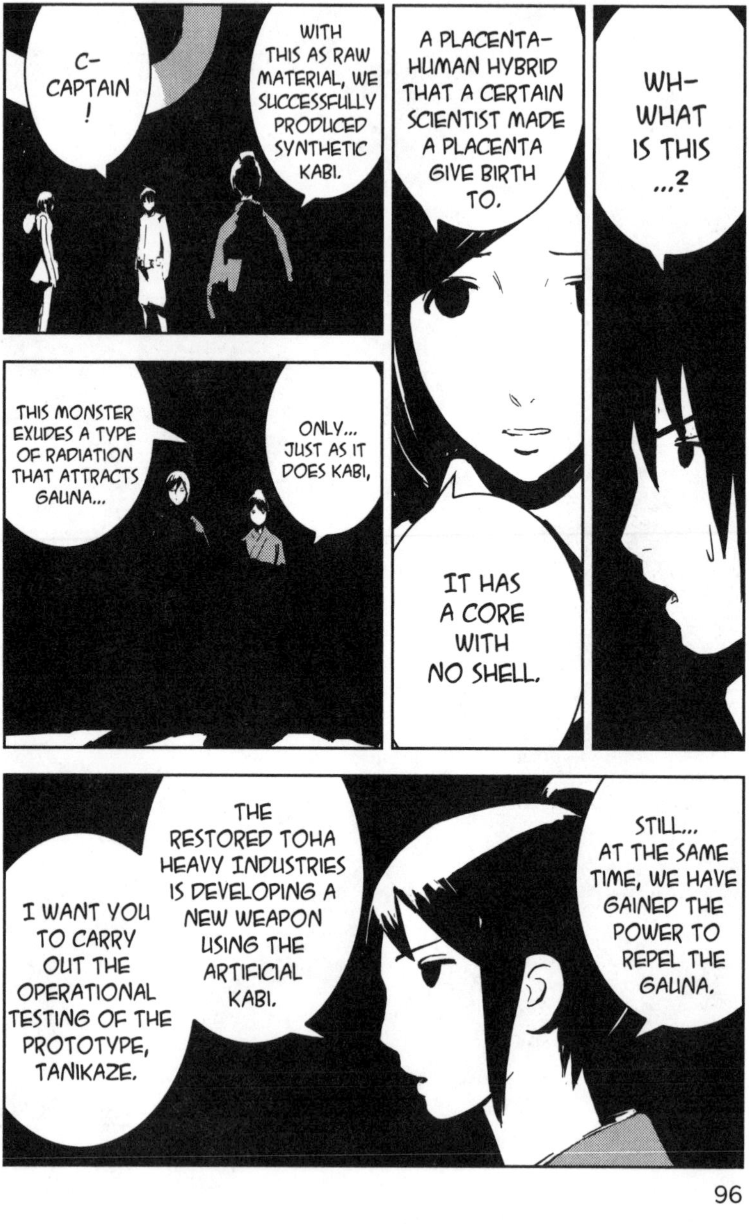
WH- WHAT IS THIS ...?
A PLACENTA-HUMAN HYBRID THAT A CERTAIN SCIENTIST MADE A PLACENTA GIVE BIRTH TO.
IT HAS A CORE WITH NO SHELL.
WITH THIS AS RAW MATERIAL, WE SUCCESSFULLY PRODUCED SYNTHETIC KABI.
C-CAPTAIN!
ONLY... JUST AS IT DOES KABI,
THIS MONSTER EXUDES A TYPE OF RADIATION THAT ATTRACTS GAUNA...
STILL... AT THE SAME TIME, WE HAVE GAINED THE POWER TO REPEL THE GAUNA.
THE RESTORED TOHA HEAVY INDUSTRIES IS DEVELOPING A NEW WEAPON USING THE ARTIFICIAL KABI.
I WANT YOU TO CARRY OUT THE OPERATIONAL TESTING OF THE PROTOTYPE, TANIKAZE.

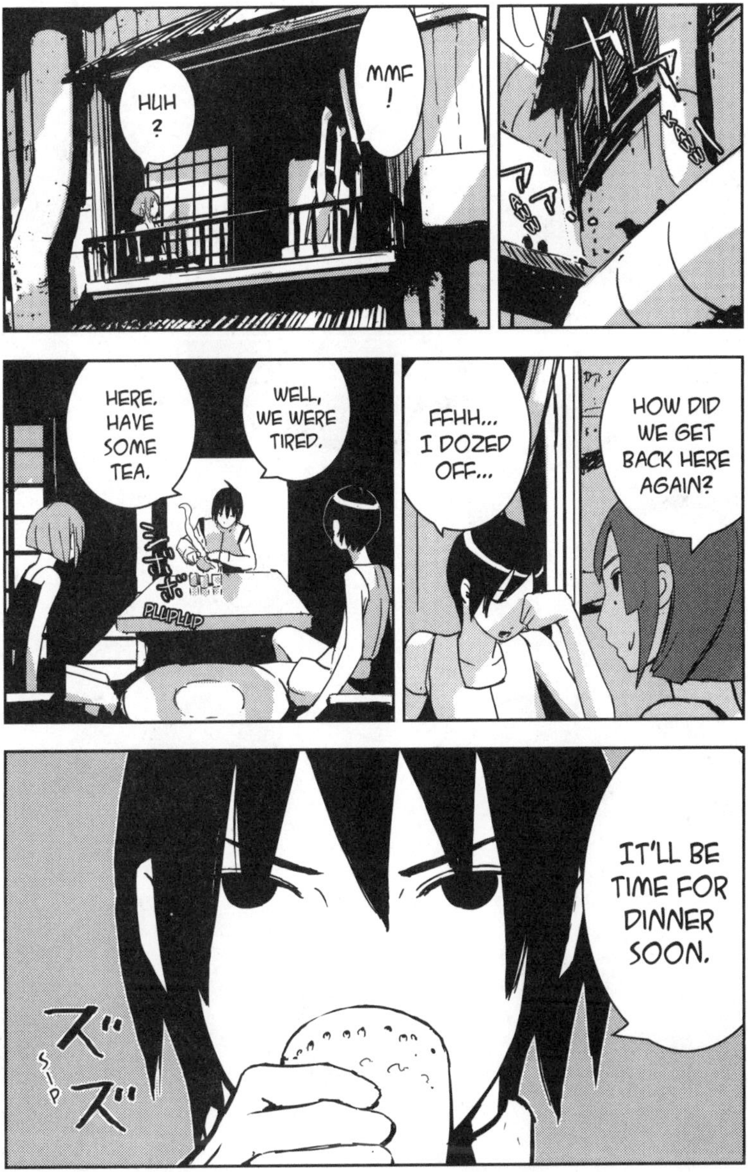
MMF !
HUH ?
HOW DID WE GET BACK HERE AGAIN?
FFHH... I DOZED OFF...
WELL, WE WERE TIRED.
HERE, HAVE SOME TEA.
PLUPLUP
IT'LL BE TIME FOR DINNER SOON.
ズズ
SIP

Chapter 18: END

Chapter 19: The Dream Izana Shinatose Had

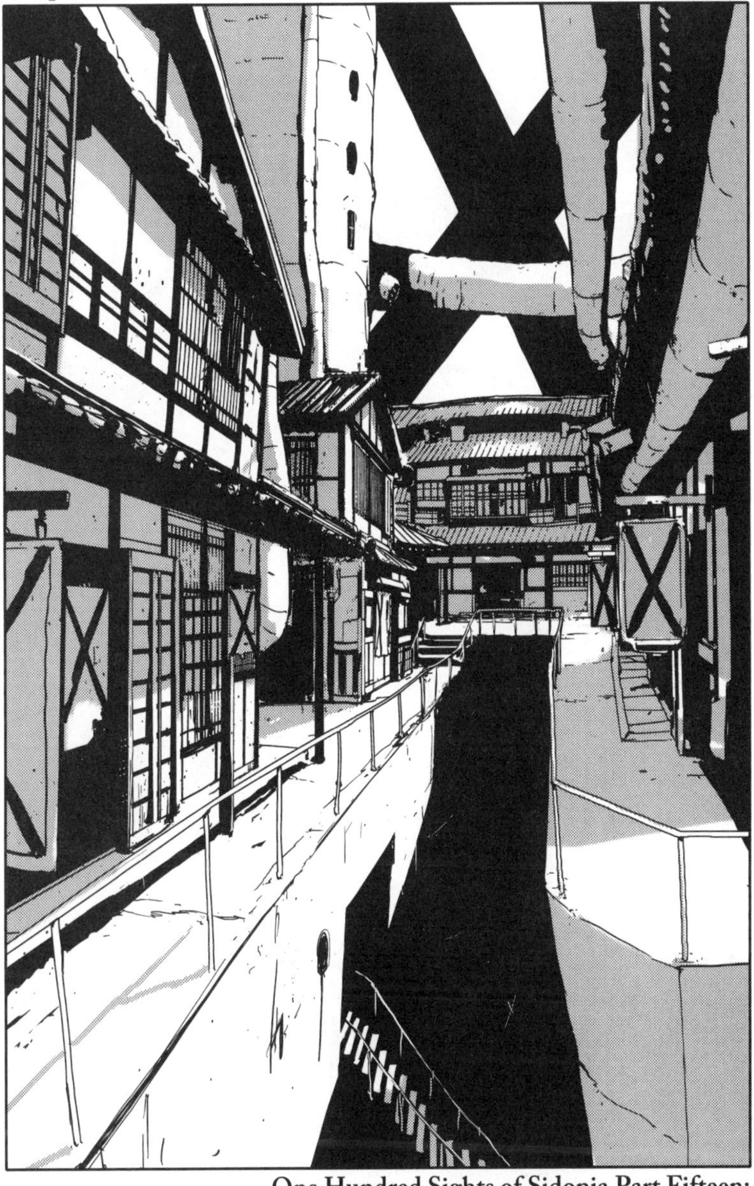

One Hundred Sights of Sidonia Part Fifteen:
Unknown (location undergoing renovations)

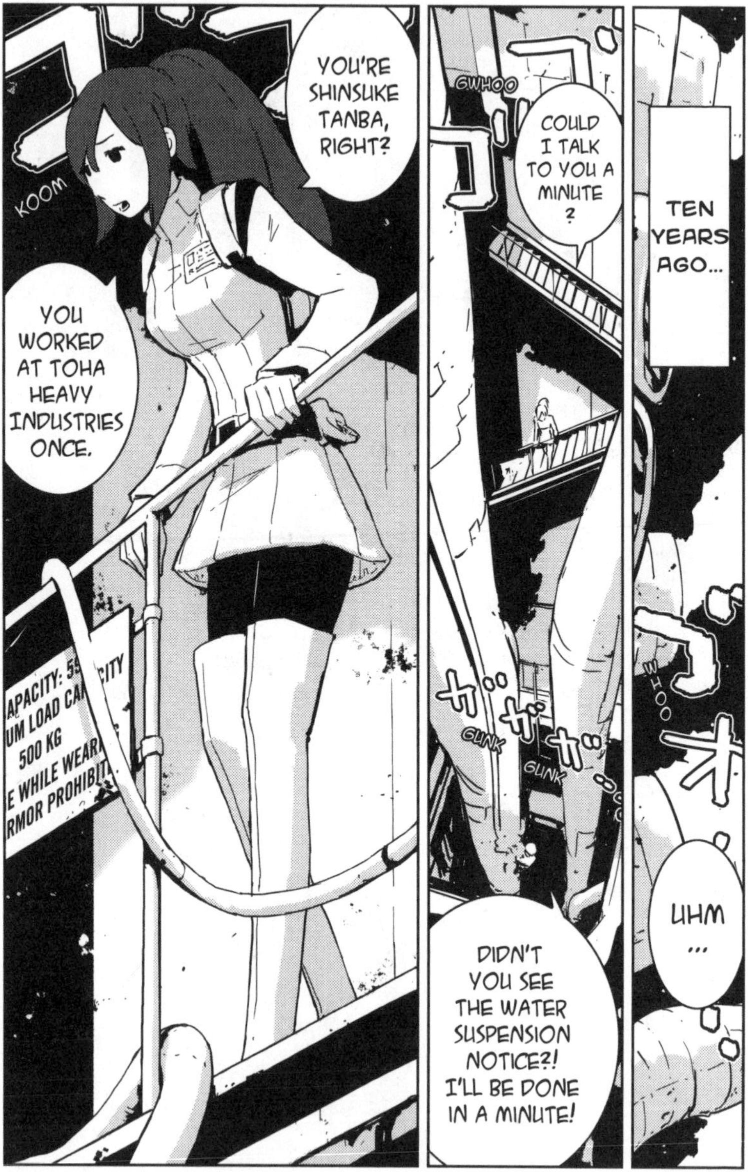
TEN YEARS AGO...
WHOO
UHM...
GWHOO
COULD I TALK TO YOU A MINUTE?
GUNK
GUNK
DIDN'T YOU SEE THE WATER SUSPENSION NOTICE?! I'LL BE DONE IN A MINUTE!
YOU'RE SHINSUKE TANBA, RIGHT?
KOOM
YOU WORKED AT TOHA HEAVY INDUSTRIES ONCE.
500 KG

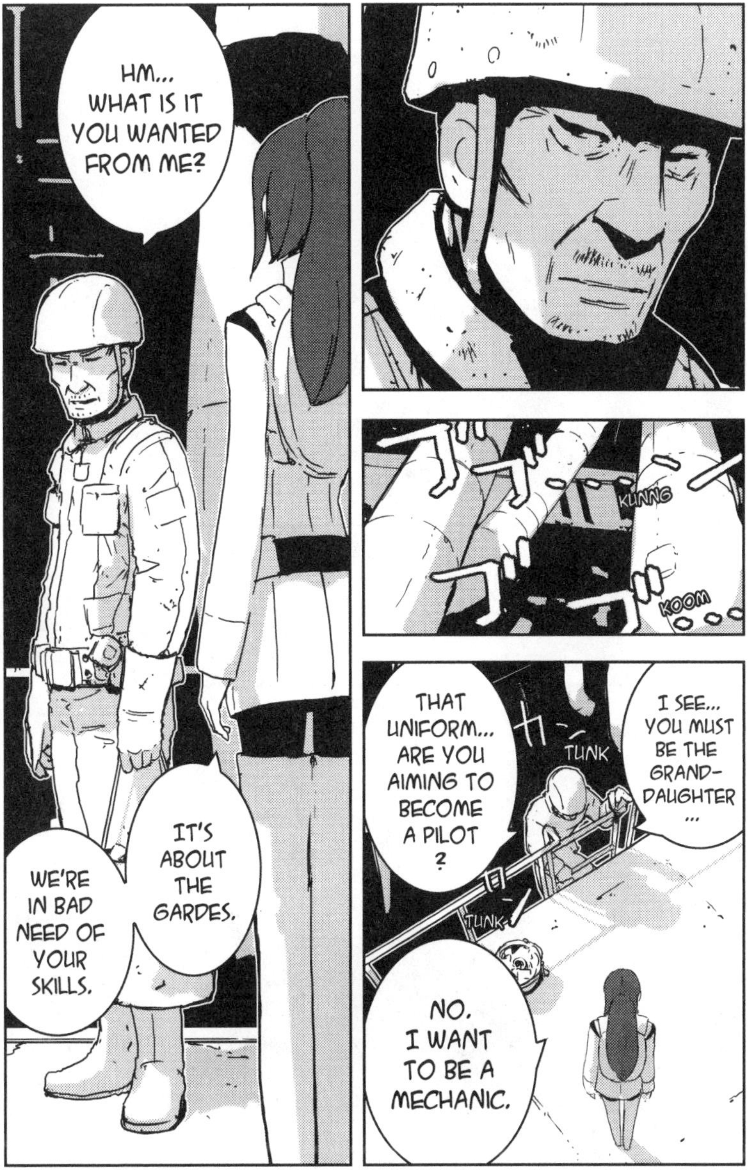
KUNNG
KOOM
I SEE... YOU MUST BE THE GRAND-DAUGHTER ...
THAT UNIFORM... ARE YOU AIMING TO BECOME A PILOT ?
TUNK
TUNK
NO. I WANT TO BE A MECHANIC.
HM... WHAT IS IT YOU WANTED FROM ME?
IT'S ABOUT THE GARDES.
WE'RE IN BAD NEED OF YOUR SKILLS.

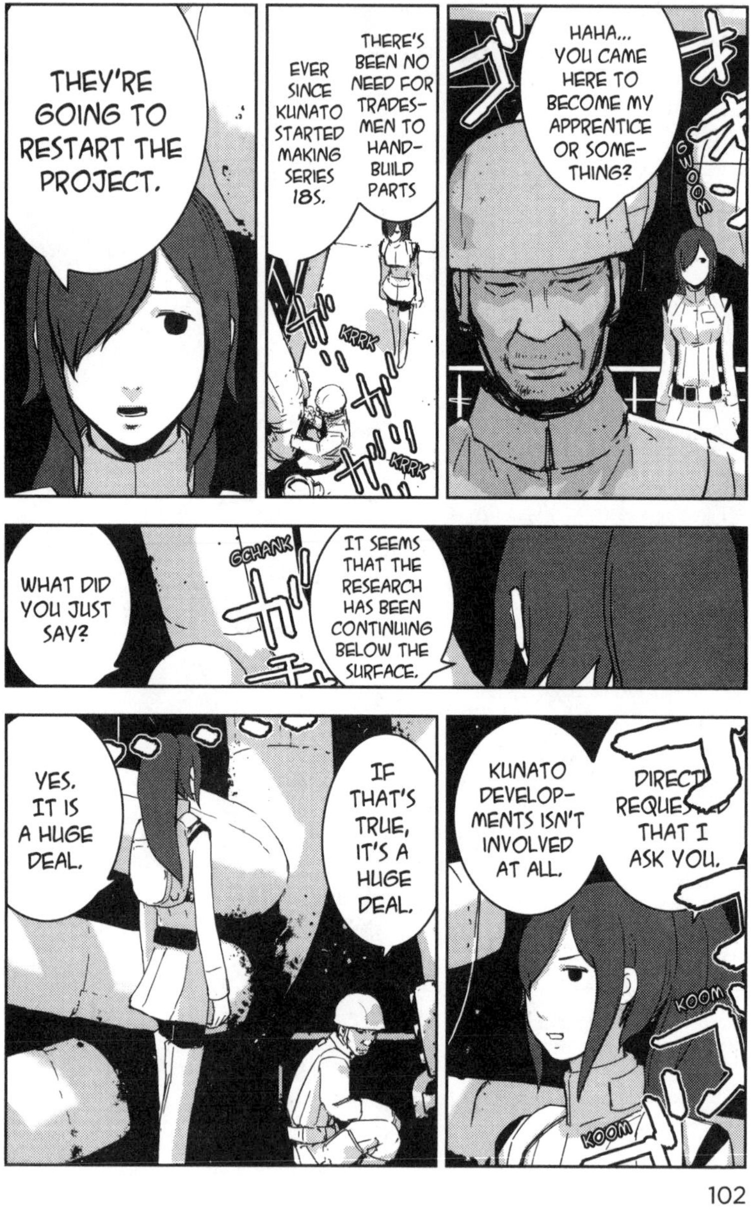
HAHA... YOU CAME HERE TO BECOME MY APPRENTICE OR SOME-THING?
WOOM
THERE'S BEEN NO NEED FOR TRADES-MEN TO HAND-BUILD PARTS
EVER SINCE KUNATO STARTED MAKING SERIES 18S.
KRRK
KRRK
THEY'RE GOING TO RESTART THE PROJECT.
IT SEEMS THAT THE RESEARCH HAS BEEN CONTINUING BELOW THE SURFACE.
GCHANK
WHAT DID YOU JUST SAY?
DIRECT REQUES THAT I ASK YOU.
KUNATO DEVELOP-MENTS ISN'T INVOLVED AT ALL.
KOOM
KOOM
IF THAT'S TRUE, IT'S A HUGE DEAL.
YES. IT IS A HUGE DEAL.

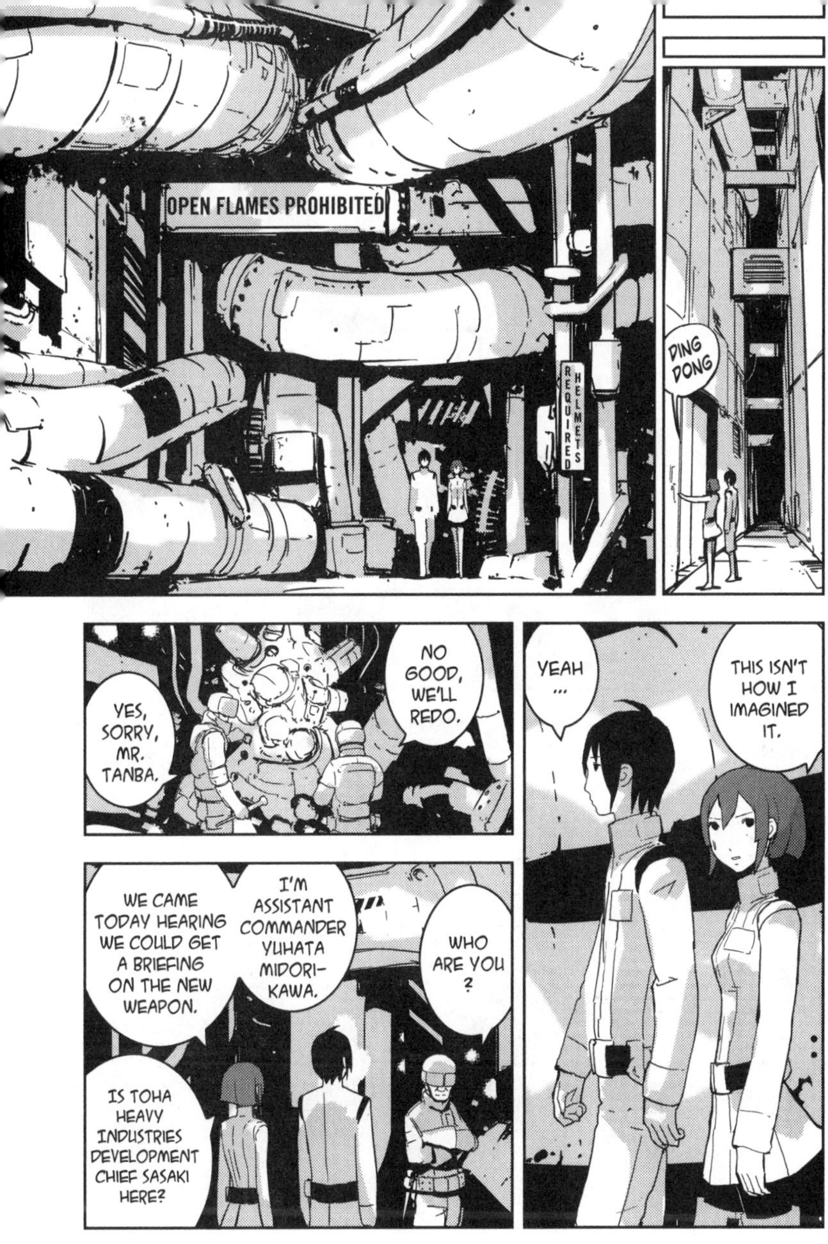
DING DONG
OPEN FLAMES PROHIBITED
HELMETS REQUIRED
THIS ISN'T HOW I IMAGINED IT.
YEAH ...
NO GOOD, WE'LL REDO.
YES, SORRY, MR. TANBA.
WHO ARE YOU ?
I'M ASSISTANT COMMANDER YUHATA MIDORI-KAWA.
WE CAME TODAY HEARING WE COULD GET A BRIEFING ON THE NEW WEAPON.
IS TOHA HEAVY INDUSTRIES DEVELOPMENT CHIEF SASAKI HERE?

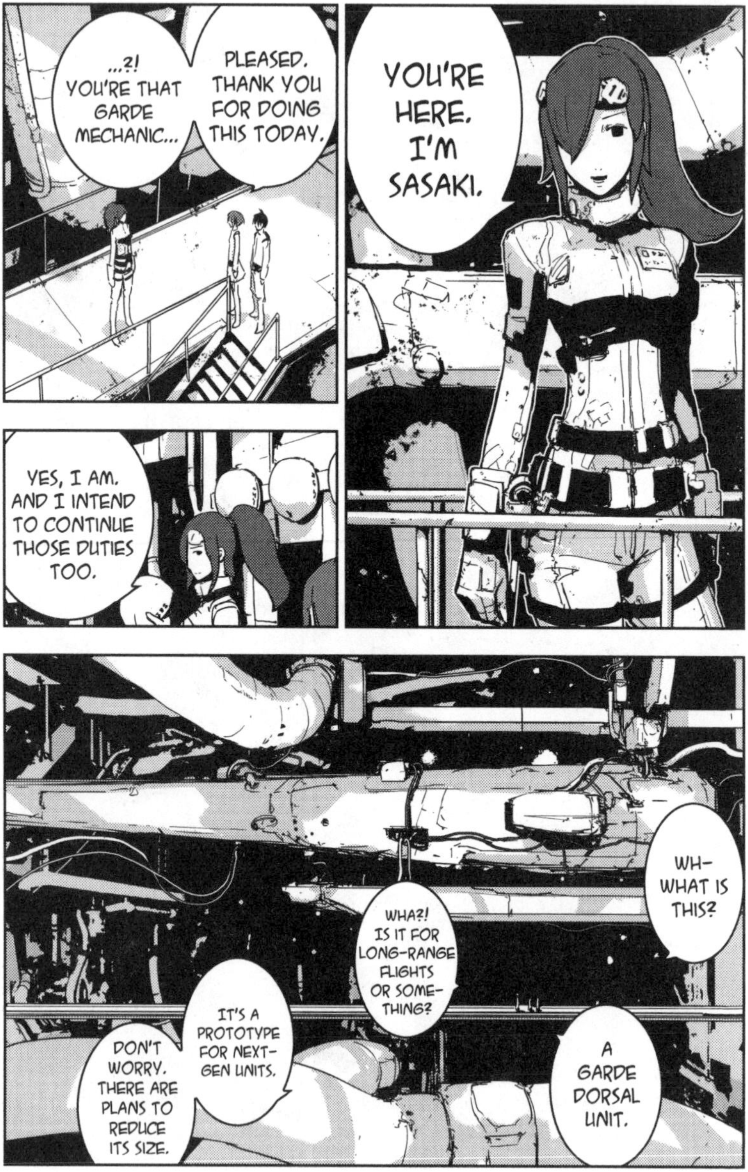
YOU'RE HERE. I'M SASAKI.
PLEASED. THANK YOU FOR DOING THIS TODAY.
...?! YOU'RE THAT GARDE MECHANIC...
YES, I AM. AND I INTEND TO CONTINUE THOSE DUTIES TOO.
WH- WHAT IS THIS?
A GARDE DORSAL UNIT.
WHA?! IS IT FOR LONG-RANGE FLIGHTS OR SOME-THING?
IT'S A PROTOTYPE FOR NEXT-GEN UNITS.
DON'T WORRY. THERE ARE PLANS TO REDUCE ITS SIZE.

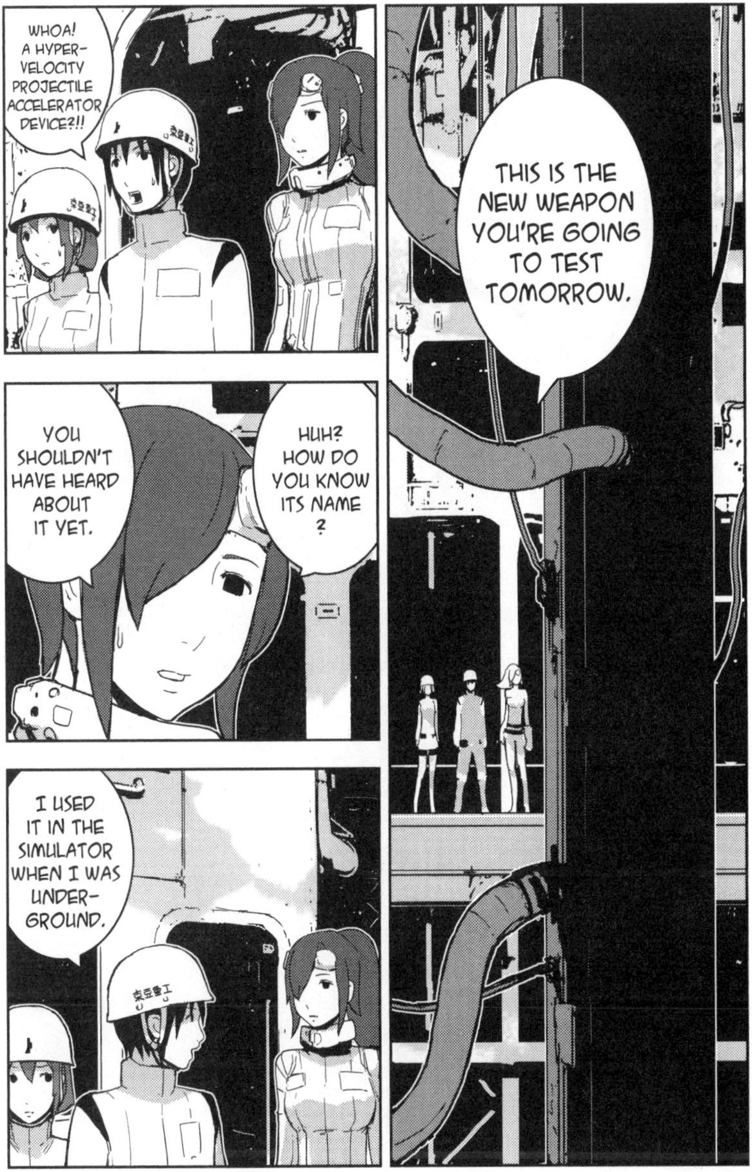
THIS IS THE NEW WEAPON YOU'RE GOING TO TEST TOMORROW.
WHOA! A HYPER-VELOCITY PROJECTILE ACCELERATOR DEVICE?!!
HUH? HOW DO YOU KNOW ITS NAME?
YOU SHOULDN'T HAVE HEARD ABOUT IT YET.
I USED IT IN THE SIMULATOR WHEN I WAS UNDER-GROUND.

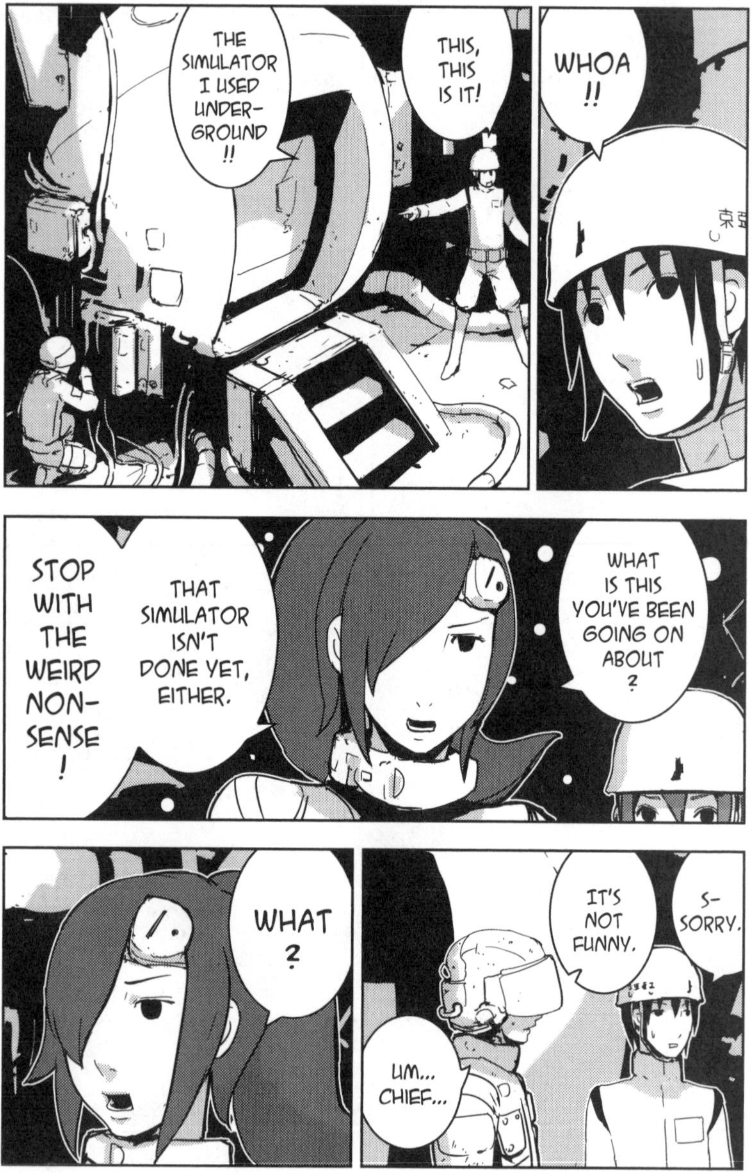
WHOA !!
THIS, THIS IS IT!
THE SIMULATOR I USED UNDERGROUND !!
WHAT IS THIS YOU'VE BEEN GOING ON ABOUT?
THAT SIMULATOR ISN'T DONE YET, EITHER.
STOP WITH THE WEIRD NONSENSE!
S-SORRY.
IT'S NOT FUNNY.
UM... CHIEF...
WHAT?

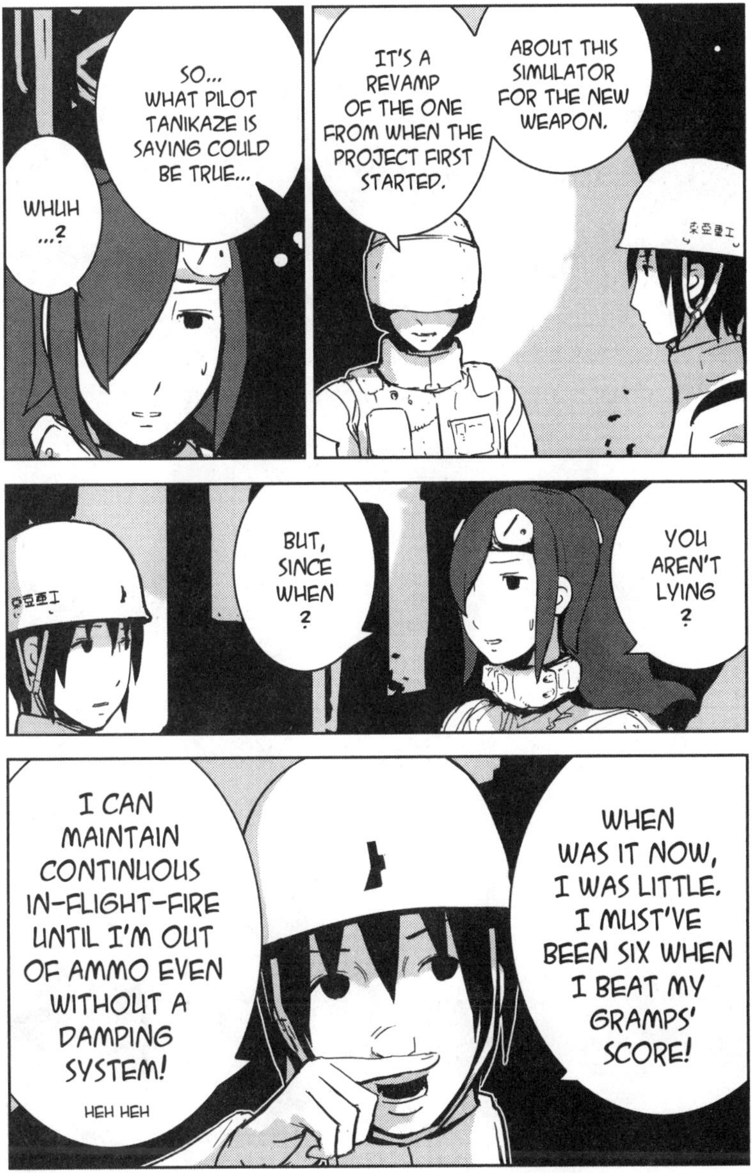
ABOUT THIS SIMULATOR FOR THE NEW WEAPON.
IT'S A REVAMP OF THE ONE FROM WHEN THE PROJECT FIRST STARTED.
SO... WHAT PILOT TANIKAZE IS SAYING COULD BE TRUE...
WHUH ...?
YOU AREN'T LYING?
BUT, SINCE WHEN?
WHEN WAS IT NOW, I WAS LITTLE. I MUST'VE BEEN SIX WHEN I BEAT MY GRAMPS' SCORE!
I CAN MAINTAIN CONTINUOUS IN-FLIGHT-FIRE UNTIL I'M OUT OF AMMO EVEN WITHOUT A DAMPING SYSTEM!
HEH HEH

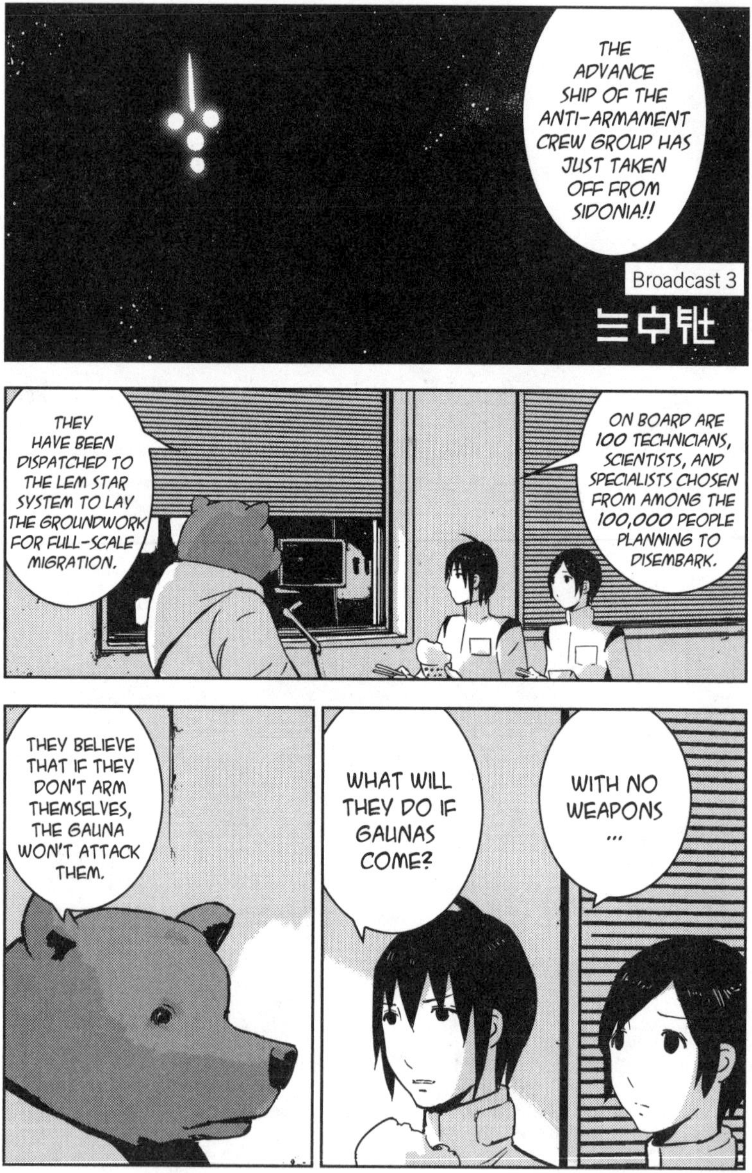
THE ADVANCE SHIP OF THE ANTI-ARMAMENT CREW GROUP HAS JUST TAKEN OFF FROM SIDONIA!!
Broadcast 3
ON BOARD ARE 100 TECHNICIANS, SCIENTISTS, AND SPECIALISTS CHOSEN FROM AMONG THE 100,000 PEOPLE PLANNING TO DISEMBARK.
THEY HAVE BEEN DISPATCHED TO THE LEM STAR SYSTEM TO LAY THE GROUNDWORK FOR FULL-SCALE MIGRATION.
WITH NO WEAPONS ...
WHAT WILL THEY DO IF GAUNAS COME?
THEY BELIEVE THAT IF THEY DON'T ARM THEMSELVES, THE GAUNA WON'T ATTACK THEM.

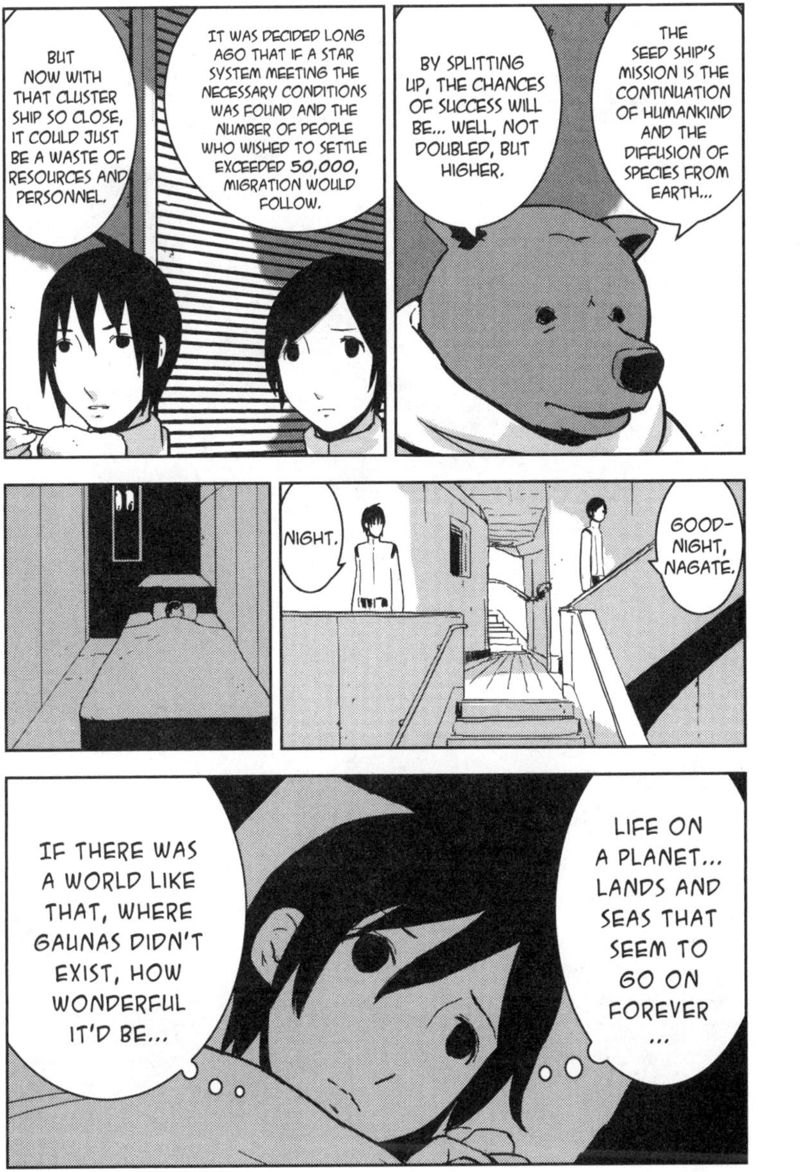
THE SEED SHIP'S MISSION IS THE CONTINUATION OF HUMANKIND AND THE DIFFUSION OF SPECIES FROM EARTH...
BY SPLITTING UP, THE CHANCES OF SUCCESS WILL BE... WELL, NOT DOUBLED, BUT HIGHER.
IT WAS DECIDED LONG AGO THAT IF A STAR SYSTEM MEETING THE NECESSARY CONDITIONS WAS FOUND AND THE NUMBER OF PEOPLE WHO WISHED TO SETTLE EXCEEDED 50,000, MIGRATION WOULD FOLLOW.
BUT NOW WITH THAT CLUSTER SHIP SO CLOSE, IT COULD JUST BE A WASTE OF RESOURCES AND PERSONNEL.
GOOD-NIGHT, NAGATE.
NIGHT.
LIFE ON A PLANET... LANDS AND SEAS THAT SEEM TO GO ON FOREVER ...
IF THERE WAS A WORLD LIKE THAT, WHERE GAUNAS DIDN'T EXIST, HOW WONDERFUL IT'D BE...

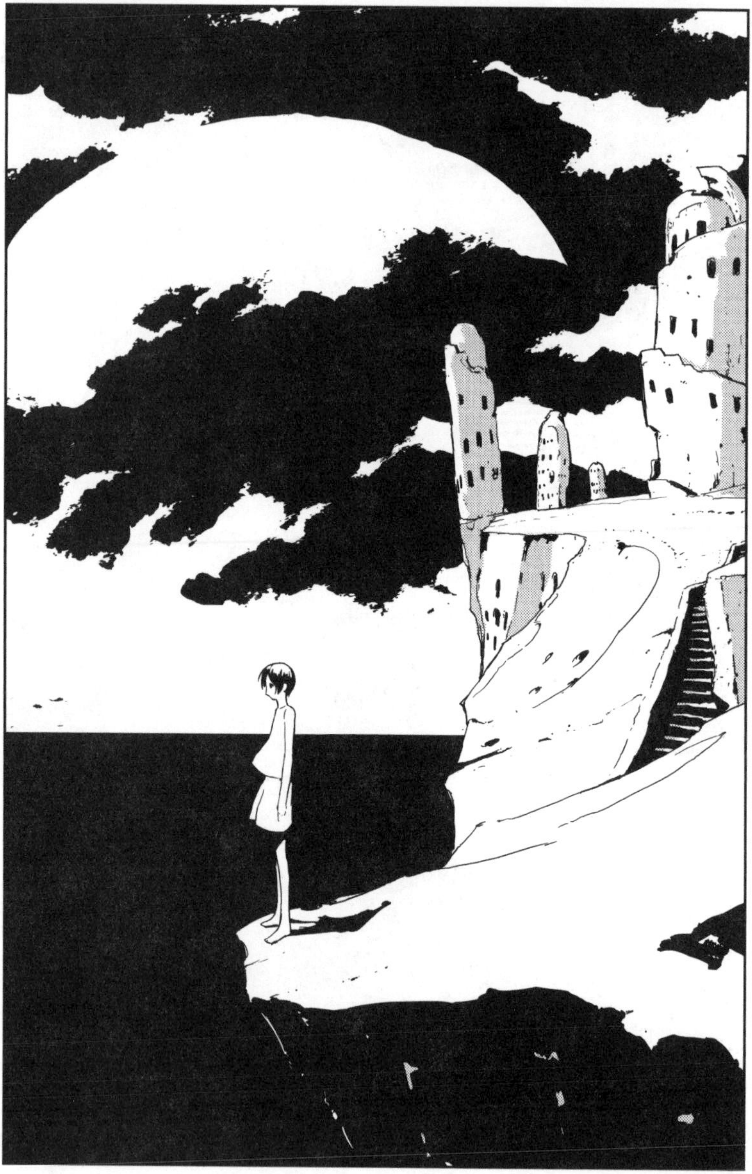

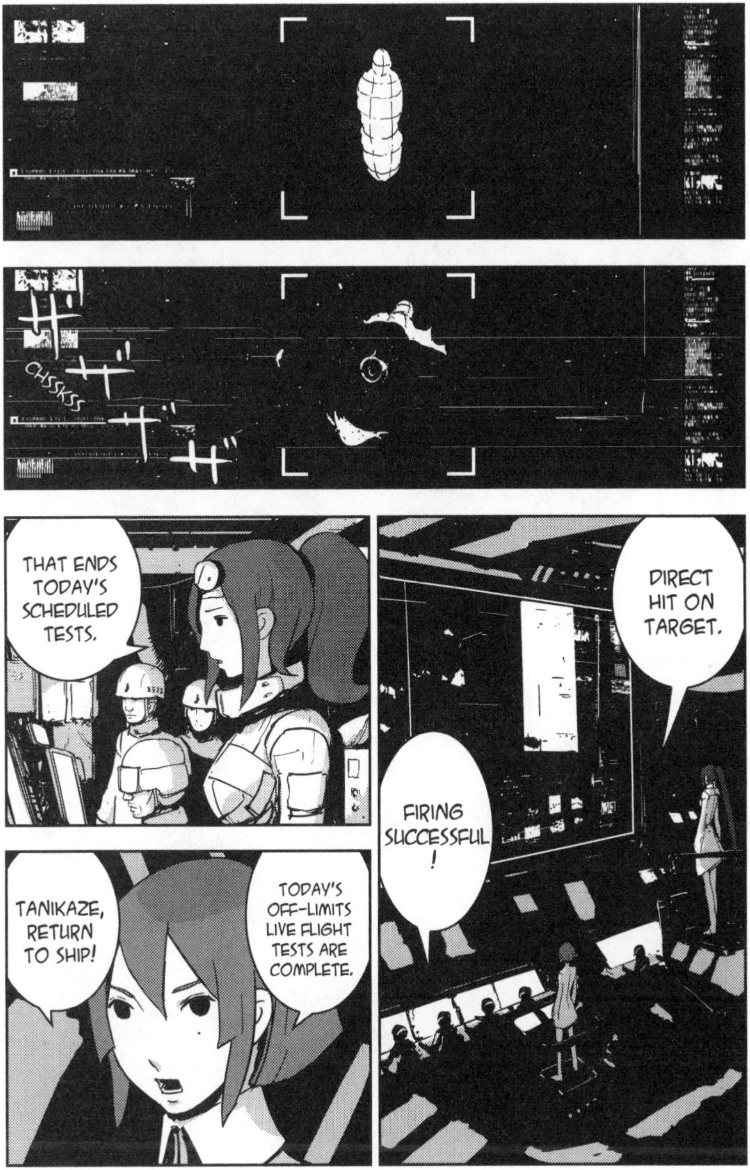
ザ ザ ザ ザ
CHSSKSS
DIRECT HIT ON TARGET.
FIRING SUCCESSFUL!
THAT ENDS TODAY'S SCHEDULED TESTS.
TODAY'S OFF-LIMITS LIVE FLIGHT TESTS ARE COMPLETE.
TANIKAZE, RETURN TO SHIP!

ROGER! 704 TANIKAZE, RETURNING TO SHIP!

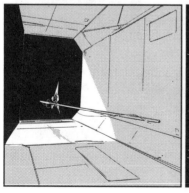

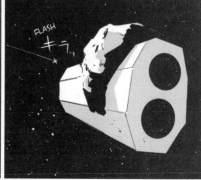

FLASH

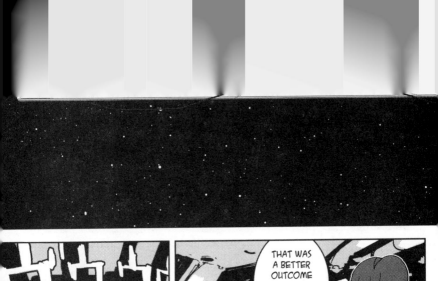

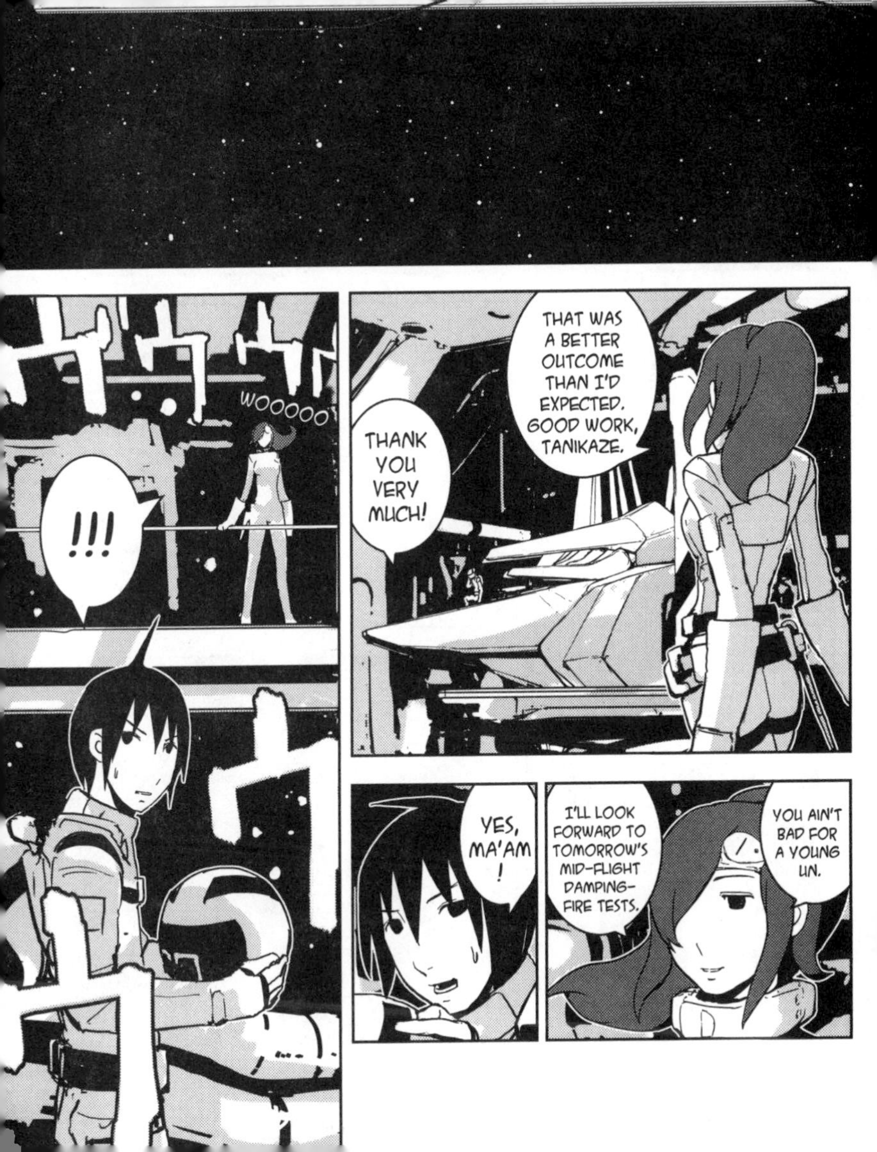
THAT WAS A BETTER OUTCOME THAN I'D EXPECTED. GOOD WORK, TANIKAZE.
THANK YOU VERY MUCH!
YOU AIN'T BAD FOR A YOUNG UN.
I'LL LOOK FORWARD TO TOMORROW'S MID-FLIGHT DAMPING-FIRE TESTS.
YES, MA'AM!
WOOOOO
!!!

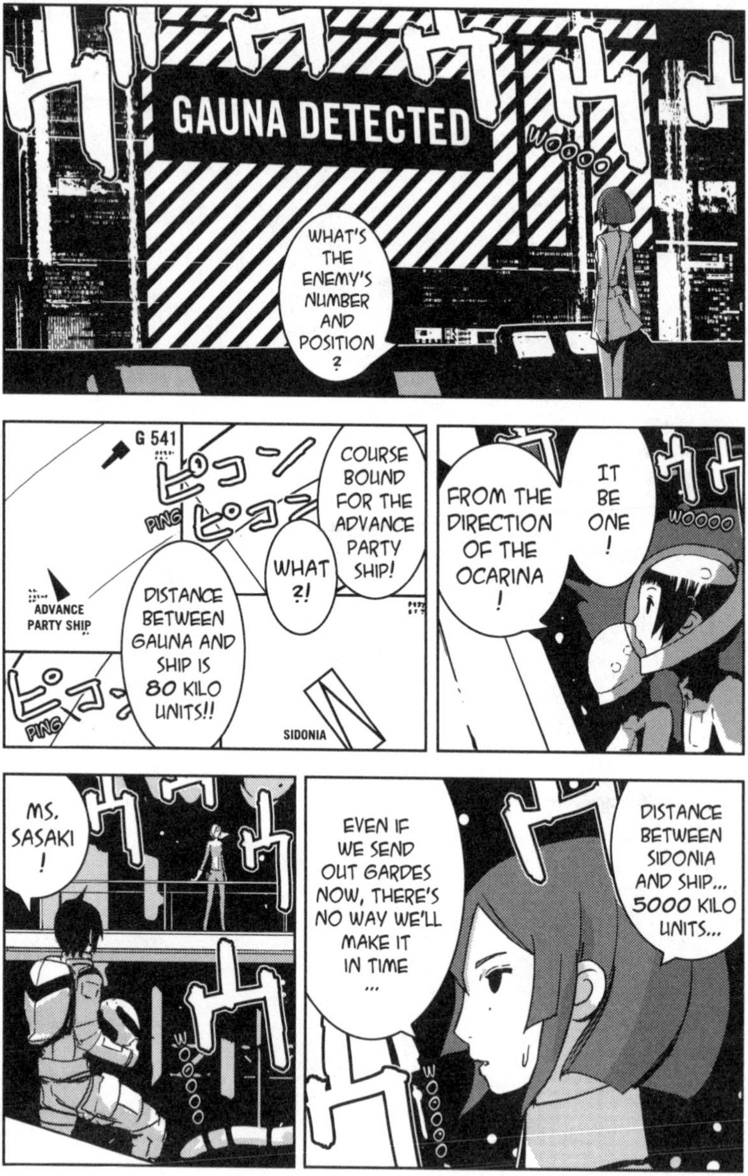
GAUNA DETECTED
WOOOO
WHAT'S THE ENEMY'S NUMBER AND POSITION?
IT BE ONE!
FROM THE DIRECTION OF THE OCARINA!
WOOOO
G 541
COURSE BOUND FOR THE ADVANCE PARTY SHIP!
WHAT?!
PING
DISTANCE BETWEEN GAUNA AND SHIP IS 80 KILO UNITS!!
ADVANCE PARTY SHIP
PING
SIDONIA
DISTANCE BETWEEN SIDONIA AND SHIP... 5000 KILO UNITS...
EVEN IF WE SEND OUT GARDES NOW, THERE'S NO WAY WE'LL MAKE IT IN TIME...
WOOOO
MS. SASAKI!
WOOOO

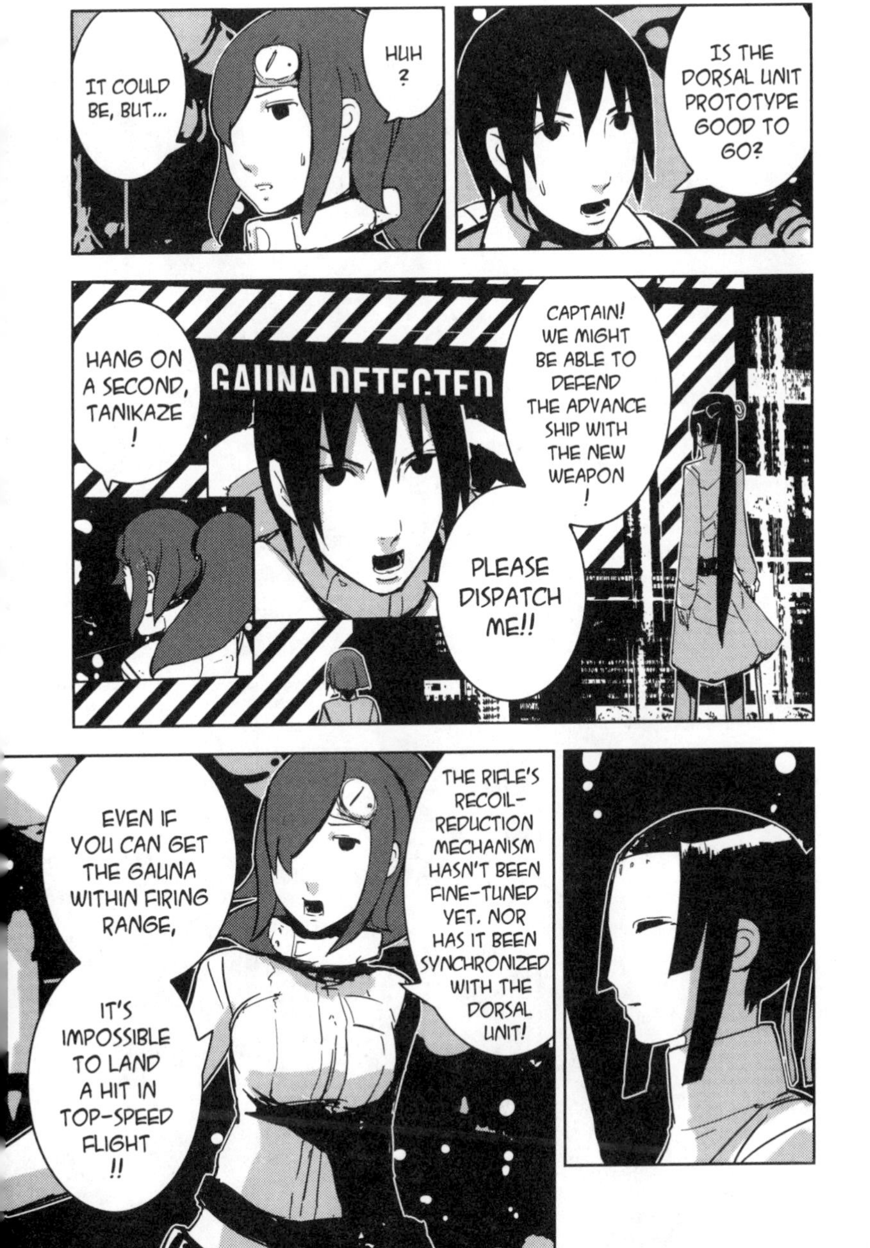
IS THE DORSAL UNIT PROTOTYPE GOOD TO GO?
HUH?
IT COULD BE, BUT...
GAUNA DETECTED
CAPTAIN! WE MIGHT BE ABLE TO DEFEND THE ADVANCE SHIP WITH THE NEW WEAPON!
PLEASE DISPATCH ME!!
HANG ON A SECOND, TANIKAZE!
THE RIFLE'S RECOIL-REDUCTION MECHANISM HASN'T BEEN FINE-TUNED YET. NOR HAS IT BEEN SYNCHRONIZED WITH THE DORSAL UNIT!
EVEN IF YOU CAN GET THE GAUNA WITHIN FIRING RANGE,
IT'S IMPOSSIBLE TO LAND A HIT IN TOP-SPEED FLIGHT!!

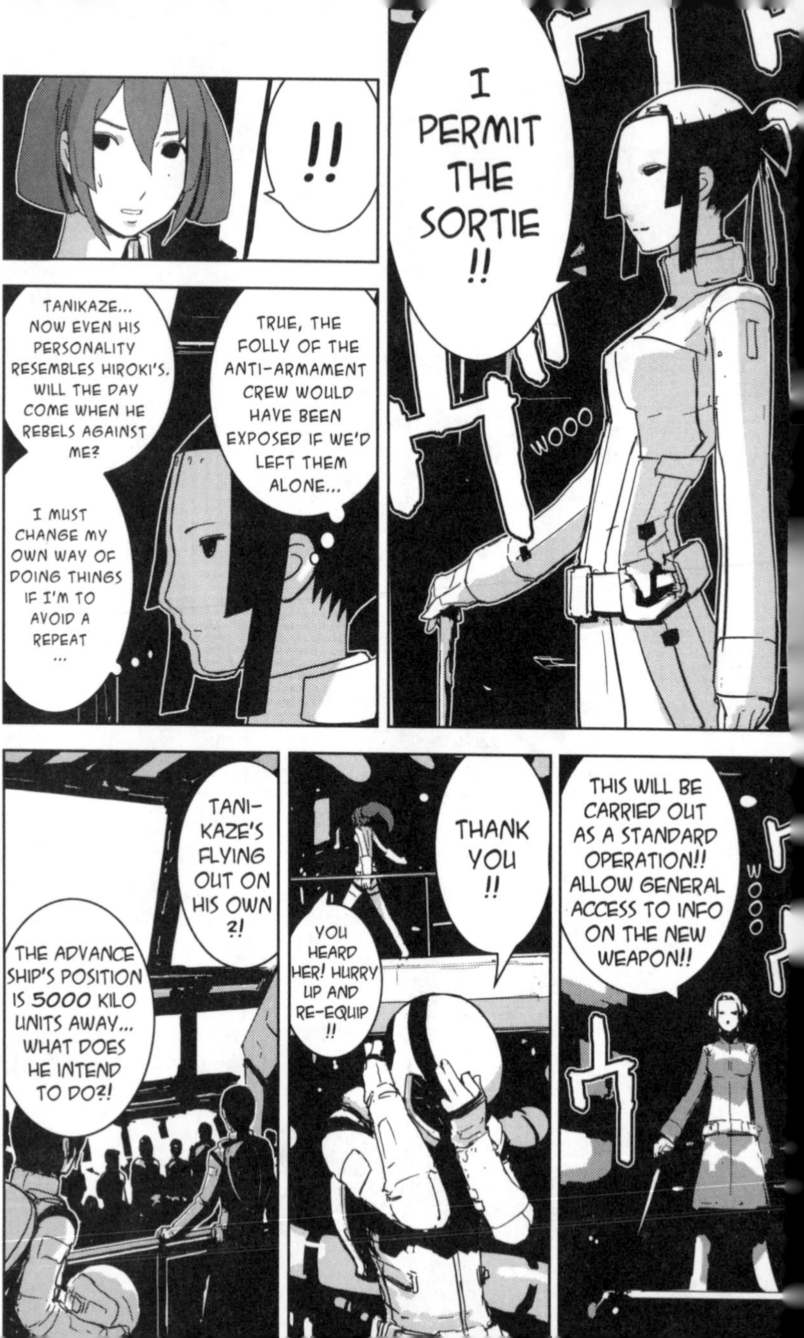

I PERMIT THE SORTIE!!
WOOO
!!
TANIKAZE... NOW EVEN HIS PERSONALITY RESEMBLES HIROKI'S. WILL THE DAY COME WHEN HE REBELS AGAINST ME?
TRUE, THE FOLLY OF THE ANTI-ARMAMENT CREW WOULD HAVE BEEN EXPOSED IF WE'D LEFT THEM ALONE...
I MUST CHANGE MY OWN WAY OF DOING THINGS IF I'M TO AVOID A REPEAT...
THIS WILL BE CARRIED OUT AS A STANDARD OPERATION!! ALLOW GENERAL ACCESS TO INFO ON THE NEW WEAPON!!
WOOO
THANK YOU!!
YOU HEARD HER! HURRY UP AND RE-EQUIP!!
TANI-KAZE'S FLYING OUT ON HIS OWN?!
THE ADVANCE SHIP'S POSITION IS 5000 KILO UNITS AWAY... WHAT DOES HE INTEND TO DO?!

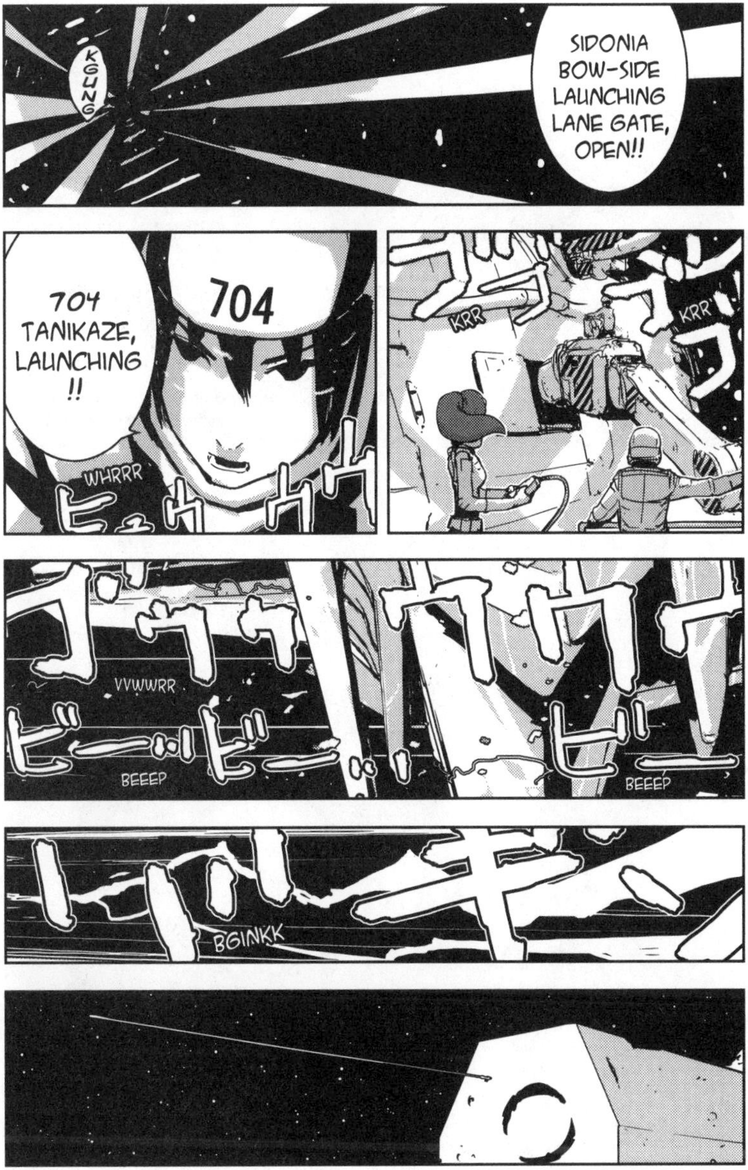
SIDONIA BOW-SIDE LAUNCHING LANE GATE, OPEN!!
KGUNG
KRR
KRR
704
704 TANIKAZE, LAUNCHING !!
WHRRR
VVWWRR
BEEEP
BEEEP
BGINKK

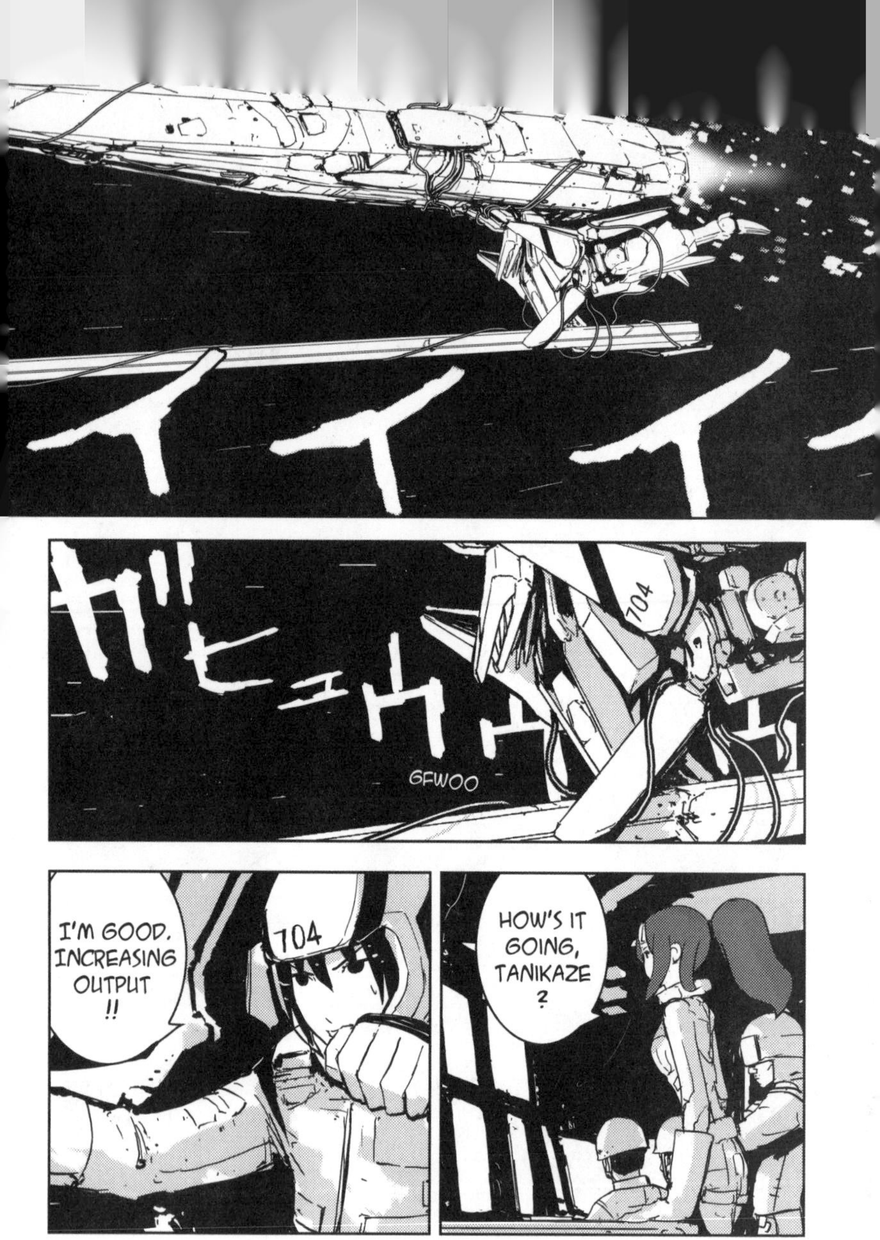

GFWOO
704
HOW'S IT GOING, TANIKAZE?
I'M GOOD. INCREASING OUTPUT!!

キイイイ
KWEEEEEM
STEP
VWM
ドン
BOOM

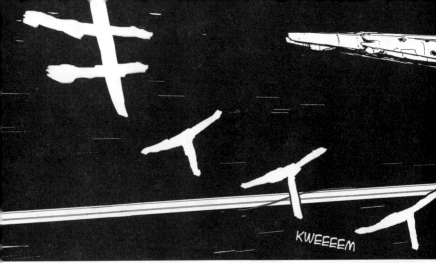

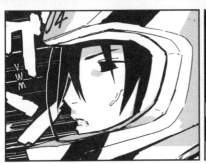

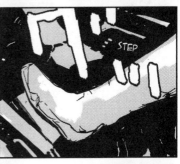

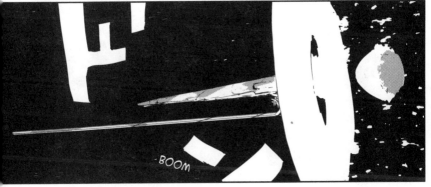

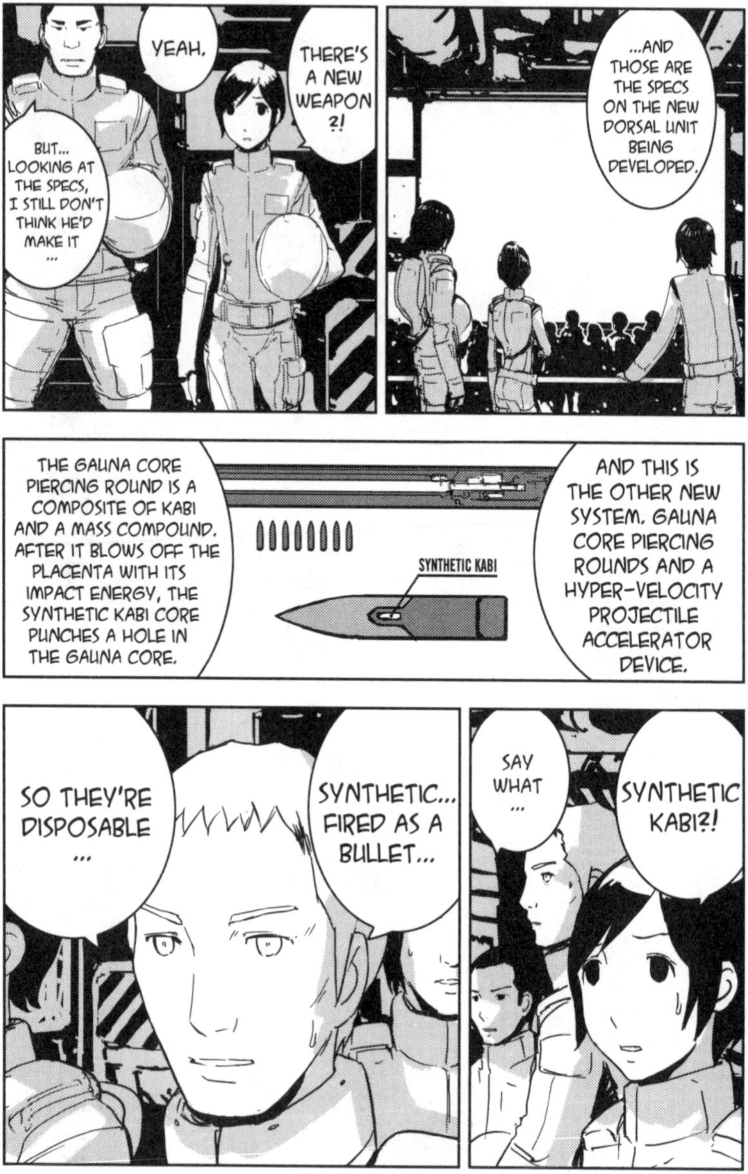
...AND THOSE ARE THE SPECS ON THE NEW DORSAL UNIT BEING DEVELOPED.
THERE'S A NEW WEAPON?!
YEAH.
BUT... LOOKING AT THE SPECS, I STILL DON'T THINK HE'D MAKE IT ...
AND THIS IS THE OTHER NEW SYSTEM. GAUNA CORE PIERCING ROUNDS AND A HYPER-VELOCITY PROJECTILE ACCELERATOR DEVICE.
SYNTHETIC KABI
THE GAUNA CORE PIERCING ROUND IS A COMPOSITE OF KABI AND A MASS COMPOUND. AFTER IT BLOWS OFF THE PLACENTA WITH ITS IMPACT ENERGY, THE SYNTHETIC KABI CORE PUNCHES A HOLE IN THE GAUNA CORE.
SYNTHETIC KABI?!
SAY WHAT ...
SYNTHETIC... FIRED AS A BULLET...
SO THEY'RE DISPOSABLE ...

GAUNA NOW LESS THAN TEN KILO UNITS FROM THE ADVANCE SHIP !!
G 541
ADVANCE PARTY SHIP
TANIKAZE UNIT IS 230 KILO UNITS FROM THE GAUNA!!
GARDE 704 TANIKAZE UNIT
THE GAUNA! IT'S ACCEL-ERATED!!
HE WON'T MAKE IT IN TIME !!
704
ENLARGED

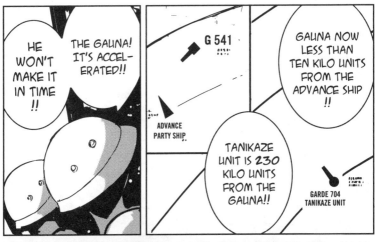

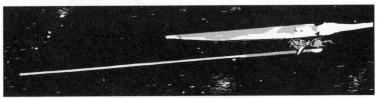

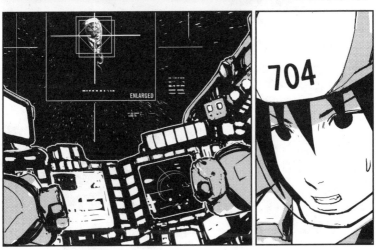

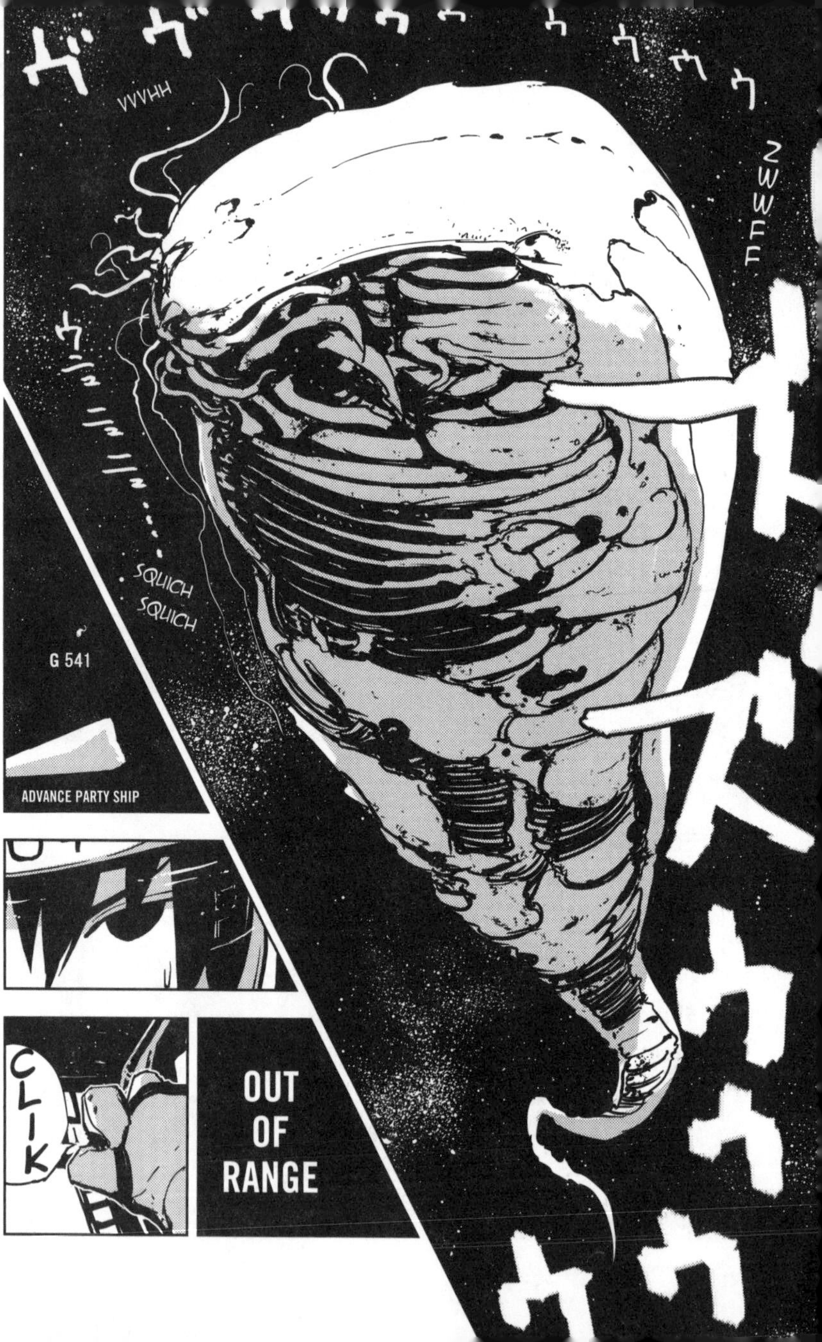

ZWWFF
VVVHH
SQUICH
SQUICH
G 541
ADVANCE PARTY SHIP
OUT OF RANGE
CLIK

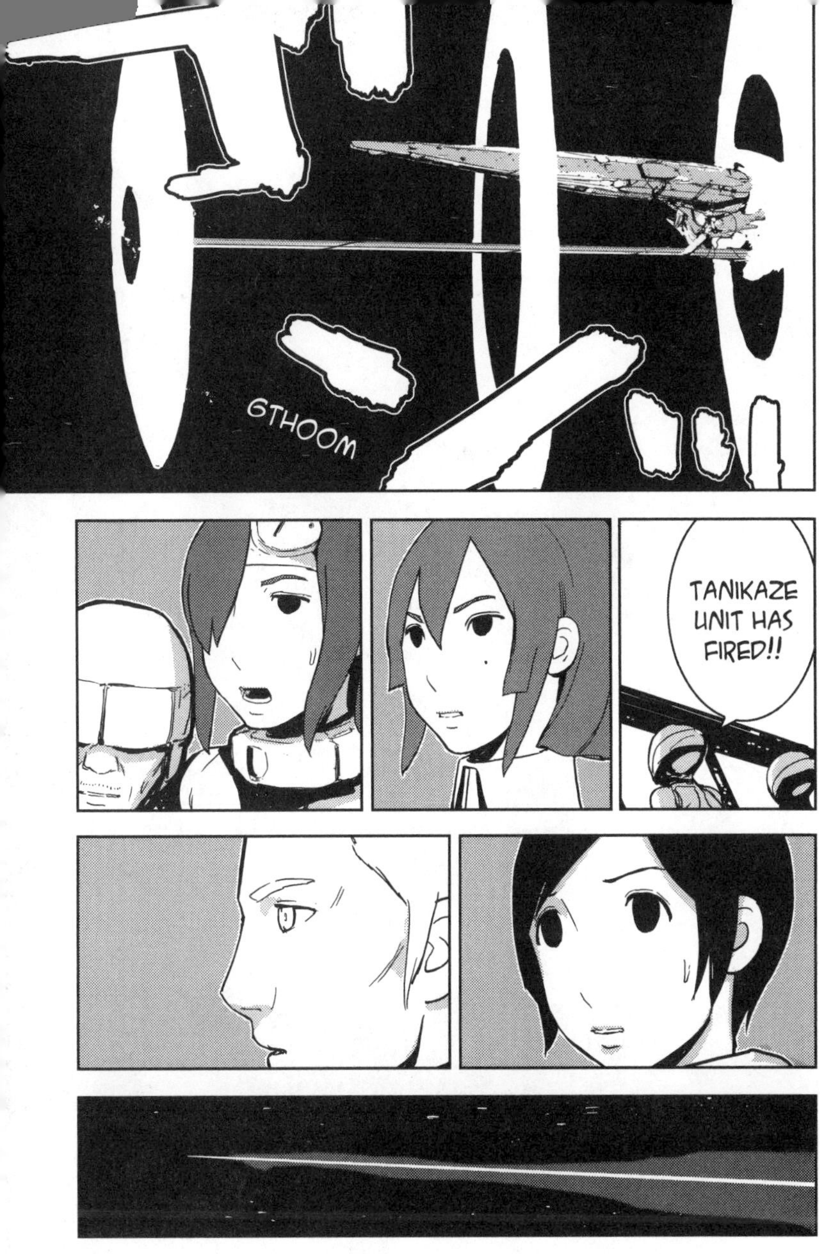
GTHOOM
TANIKAZE UNIT HAS FIRED!!

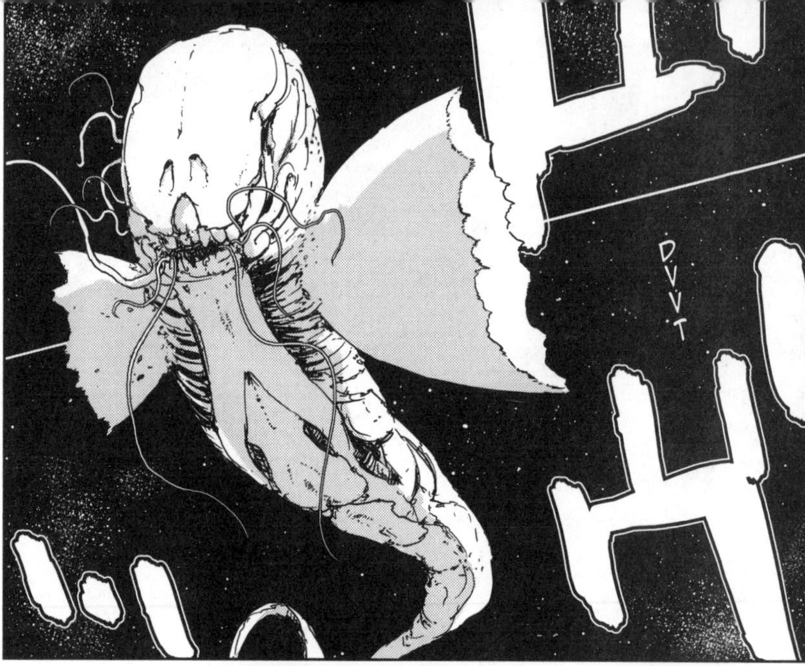
DVVT

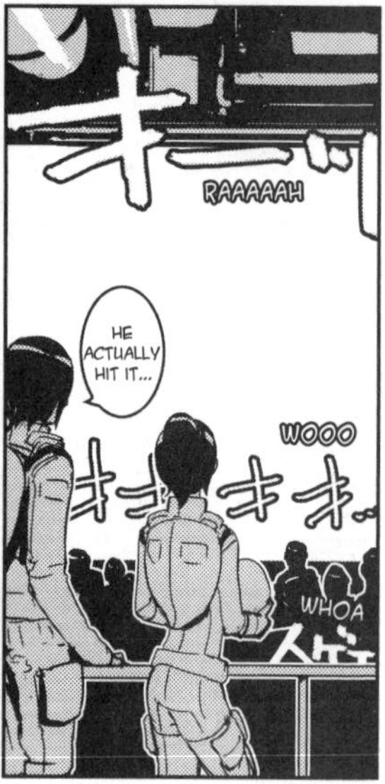
RAAAAAH
HE ACTUALLY HIT IT...
WOOO
WHOA

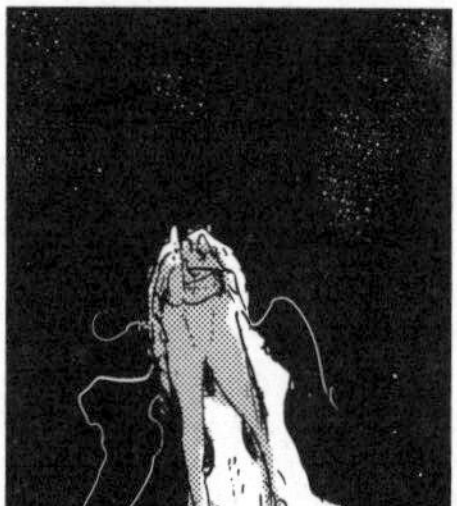
GWOOO

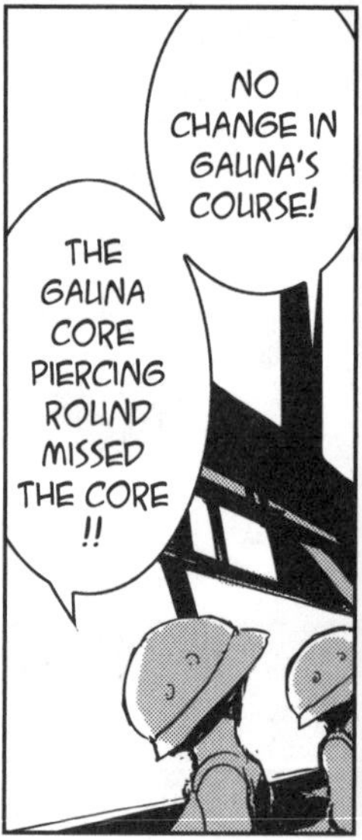
NO CHANGE IN GAUNA'S COURSE!
THE GAUNA CORE PIERCING ROUND MISSED THE CORE !!

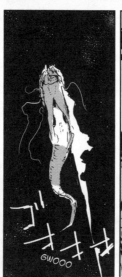

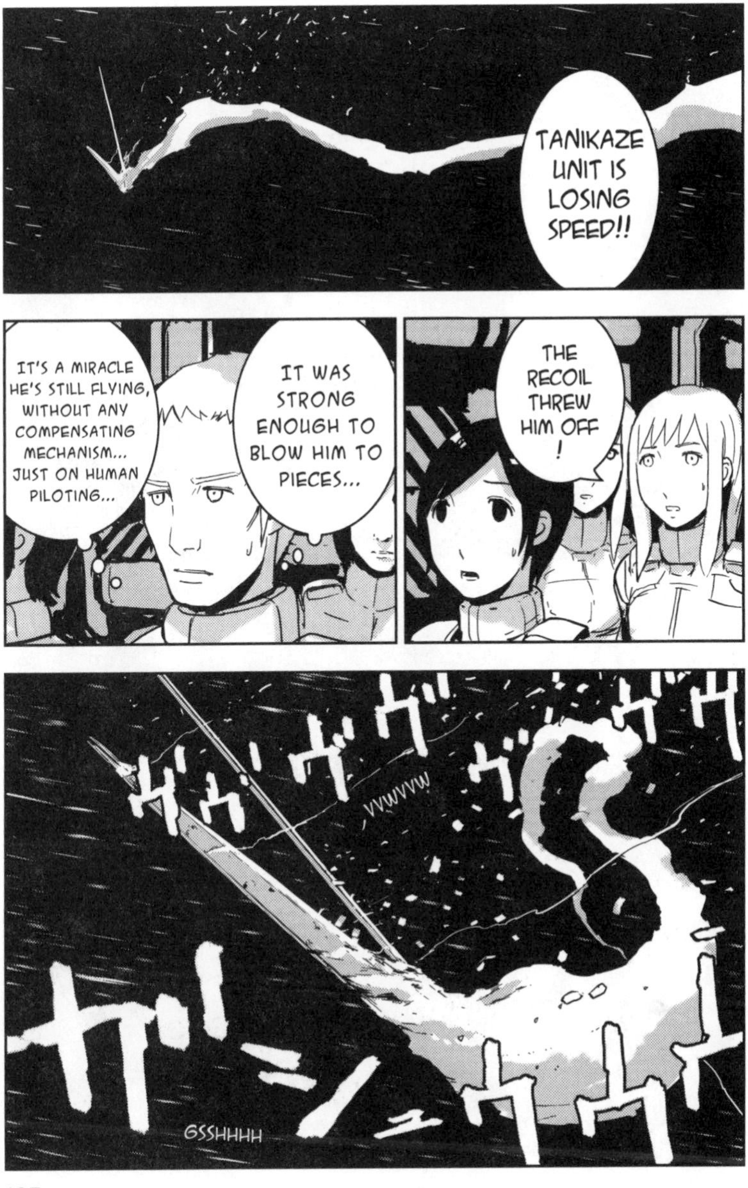
TANIKAZE UNIT IS LOSING SPEED!!
THE RECOIL THREW HIM OFF!
IT WAS STRONG ENOUGH TO BLOW HIM TO PIECES...
IT'S A MIRACLE HE'S STILL FLYING, WITHOUT ANY COMPENSATING MECHANISM... JUST ON HUMAN PILOTING...
VVWVVW
GSSHHHH

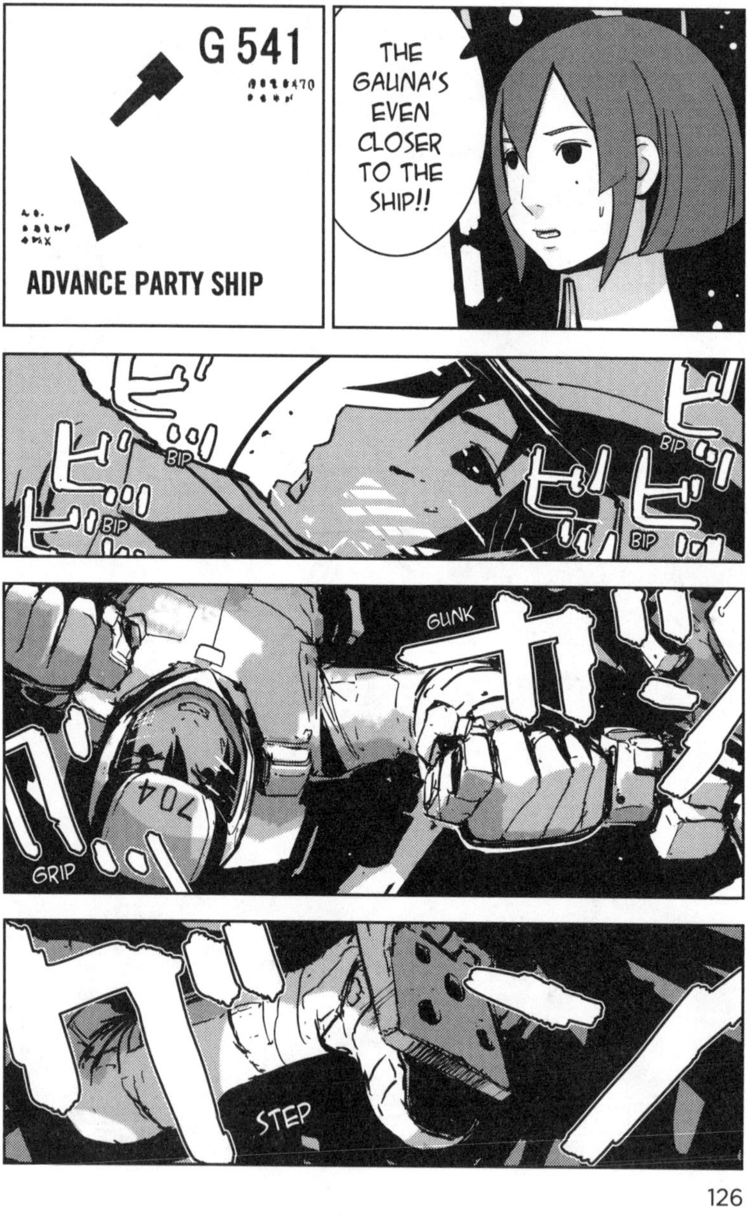
THE GAUNA'S EVEN CLOSER TO THE SHIP!!
G 541
ADVANCE PARTY SHIP
BIP
BIP
BIP
BIP
GUNK
704
GRIP
STEP

BWOOF
HE'S RIGHTED HIMSELF !!
BIP
BIP
BIP
BIP
BIP
カチ
CLIK
704

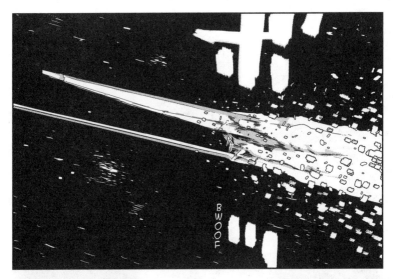

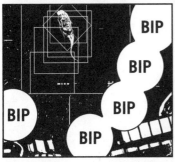

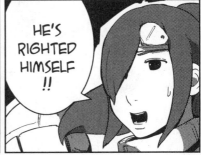

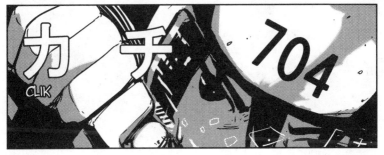

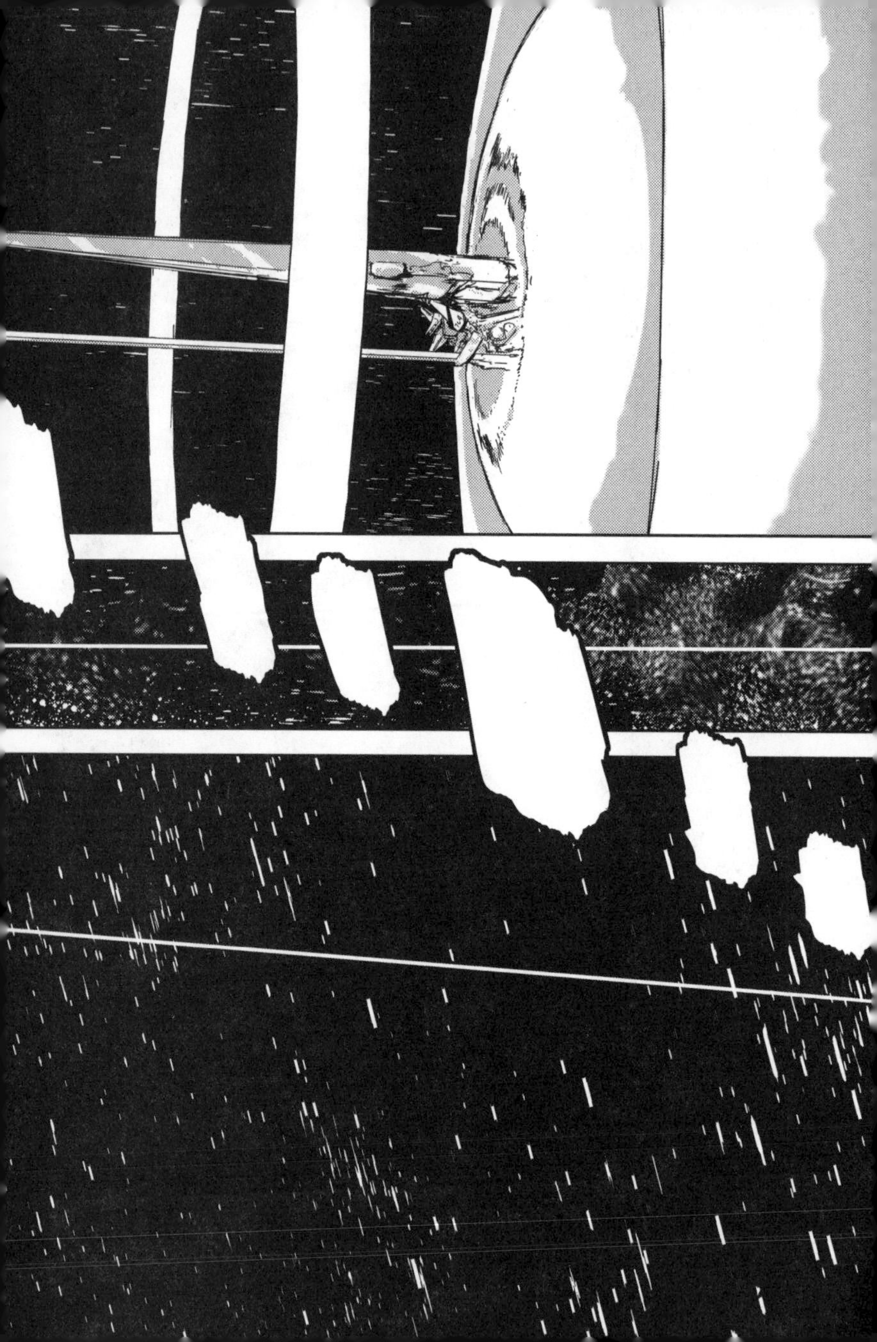

DOMM

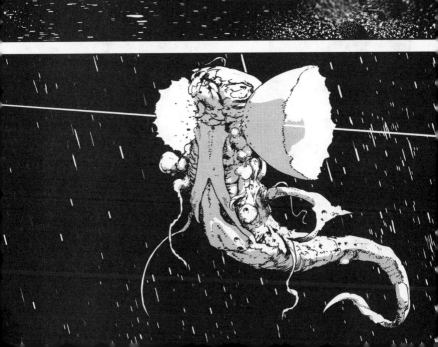

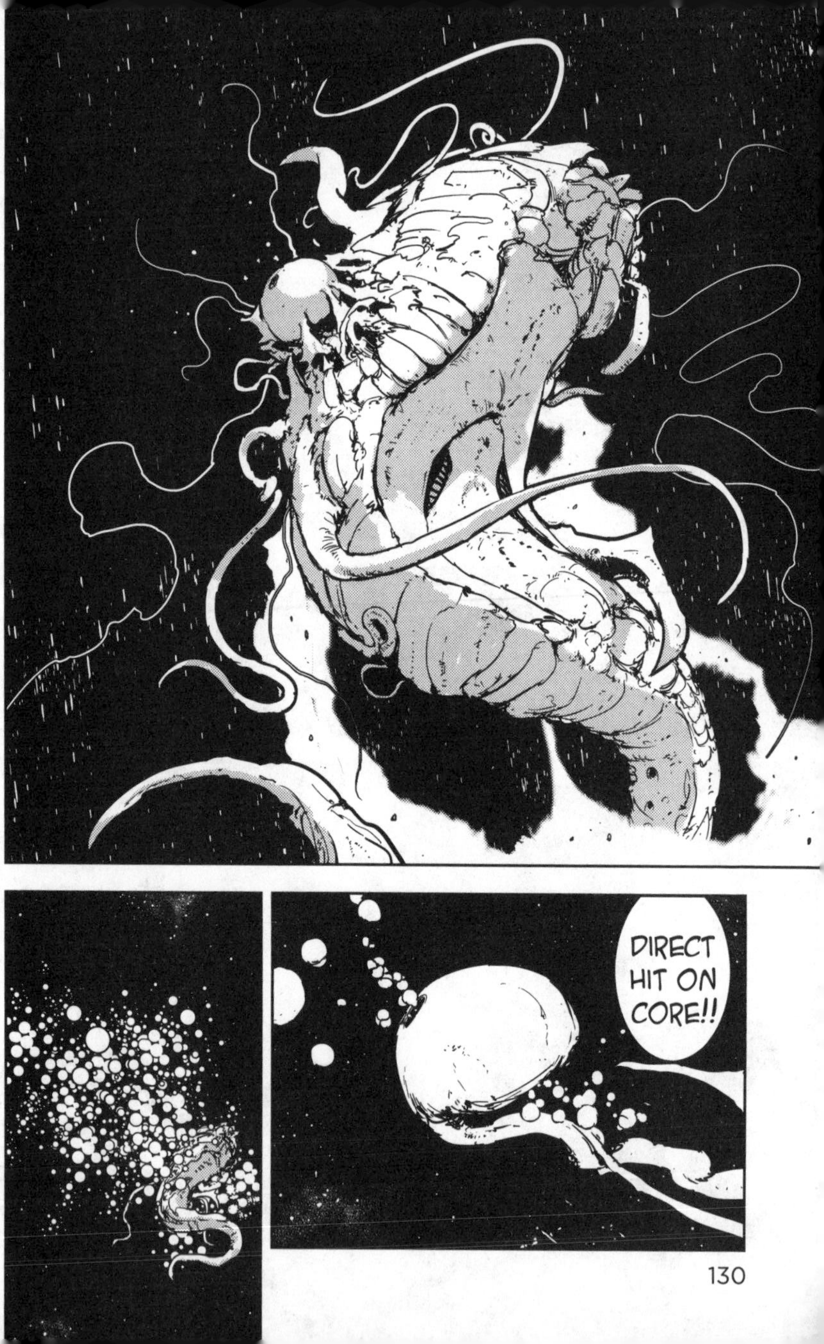
DIRECT HIT ON CORE!!

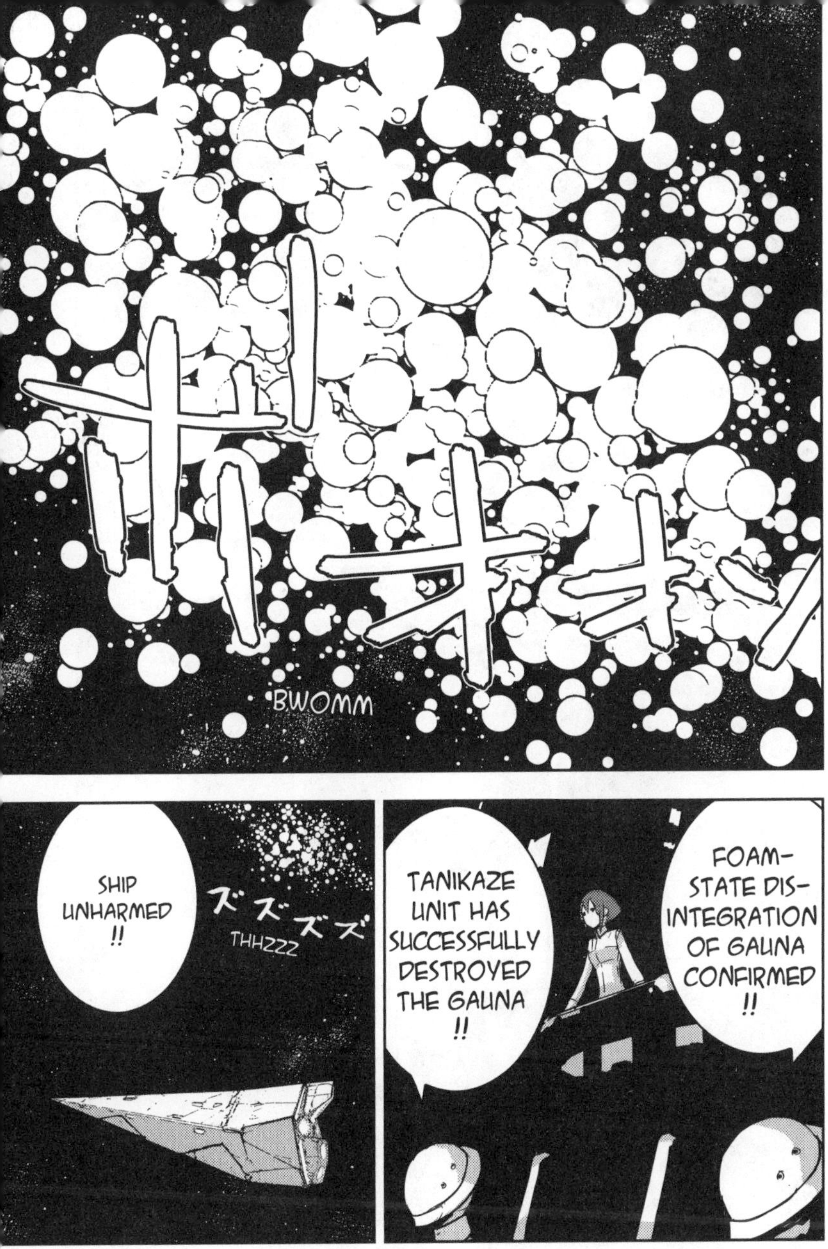
BWOMM
FOAM-STATE DIS-INTEGRATION OF GAUNA CONFIRMED !!
TANIKAZE UNIT HAS SUCCESSFULLY DESTROYED THE GAUNA !!
SHIP UNHARMED !!
ズズズズ
THHZZZ

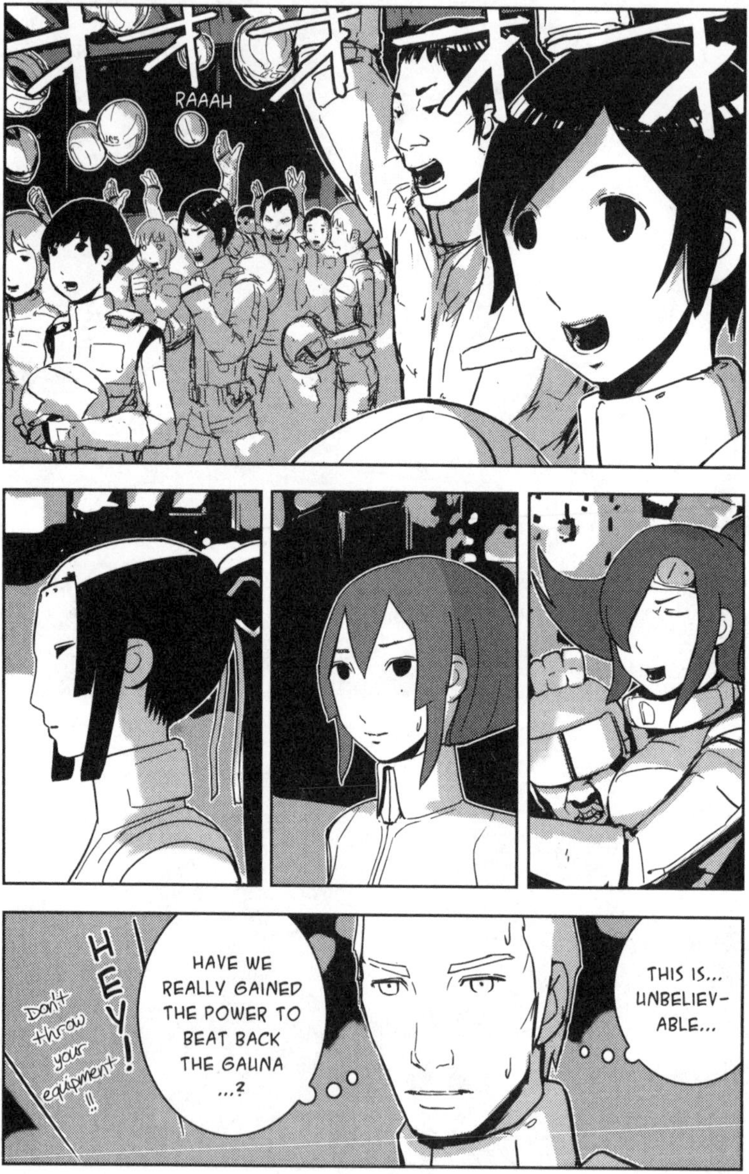
オオオーオ
RAAAH
THIS IS... UNBELIEV-ABLE...
HAVE WE REALLY GAINED THE POWER TO BEAT BACK THE GAUNA ...?
HEY!
Don't throw your equipment!!

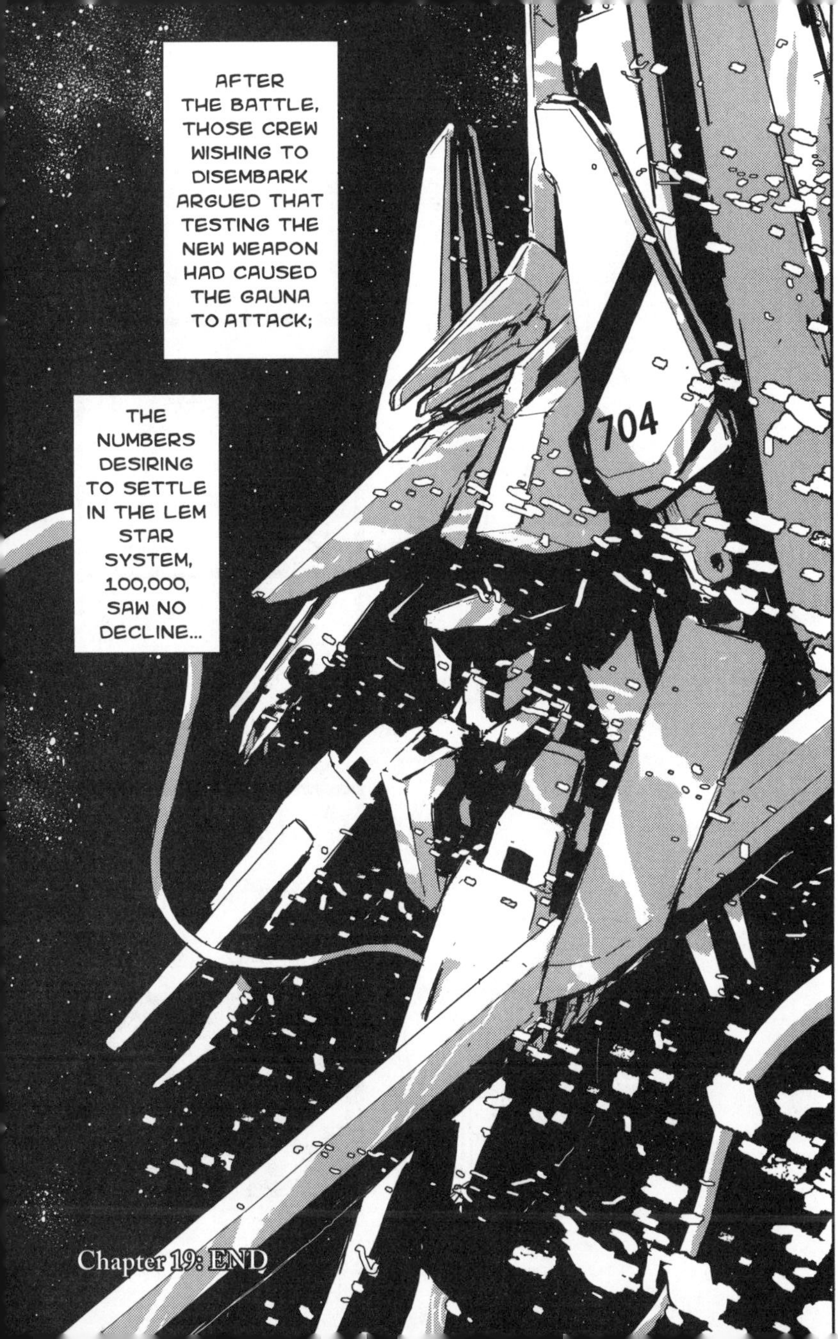
AFTER THE BATTLE, THOSE CREW WISHING TO DISEMBARK ARGUED THAT TESTING THE NEW WEAPON HAD CAUSED THE GAUNA TO ATTACK;
THE NUMBERS DESIRING TO SETTLE IN THE LEM STAR SYSTEM, 100,000, SAW NO DECLINE...
704
Chapter 19: END

Cluster Ship
Interior Tour
WHRRRLL
HSSHH
SHH
ガギガギガギ
GJNG
GJNG
GJNG
GCHING
BLUBLUBLUB
GLUBLUBLUB
HWRGGG

G78

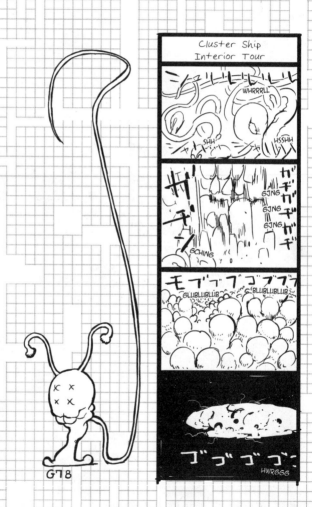

Chapter 20: Norio Kunato's Turmoil

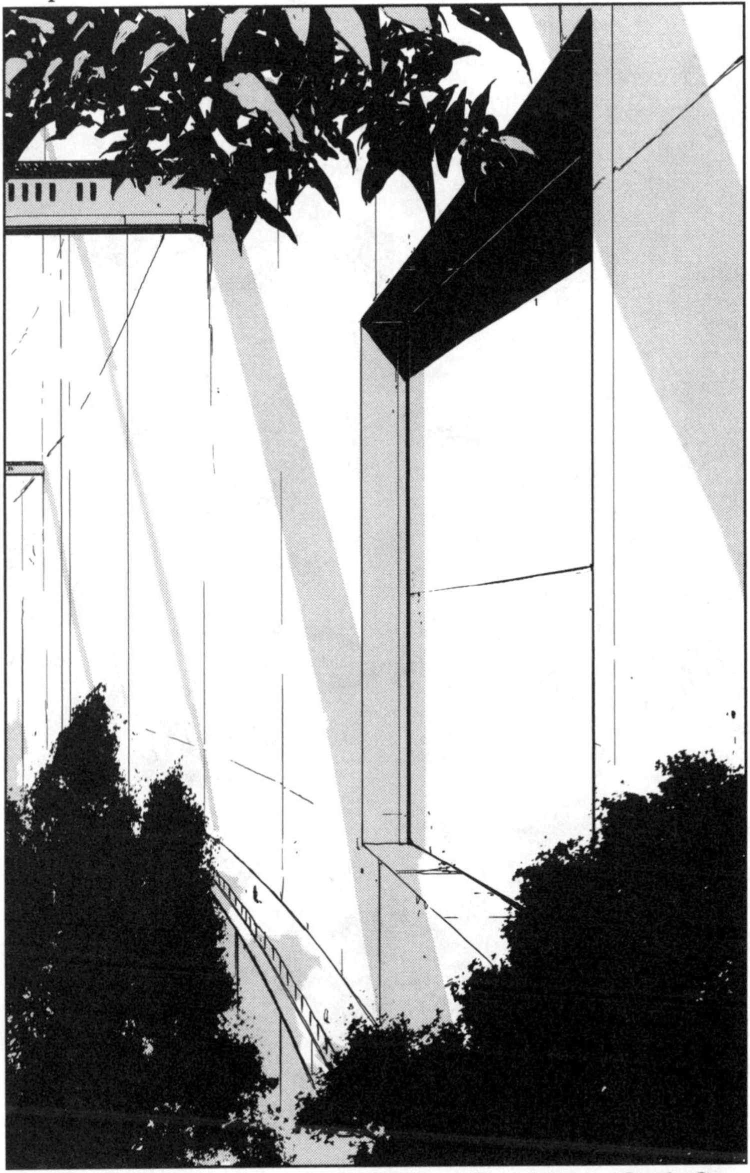

One Hundred Sights of Sidonia Part Sixteen: Garde Gate

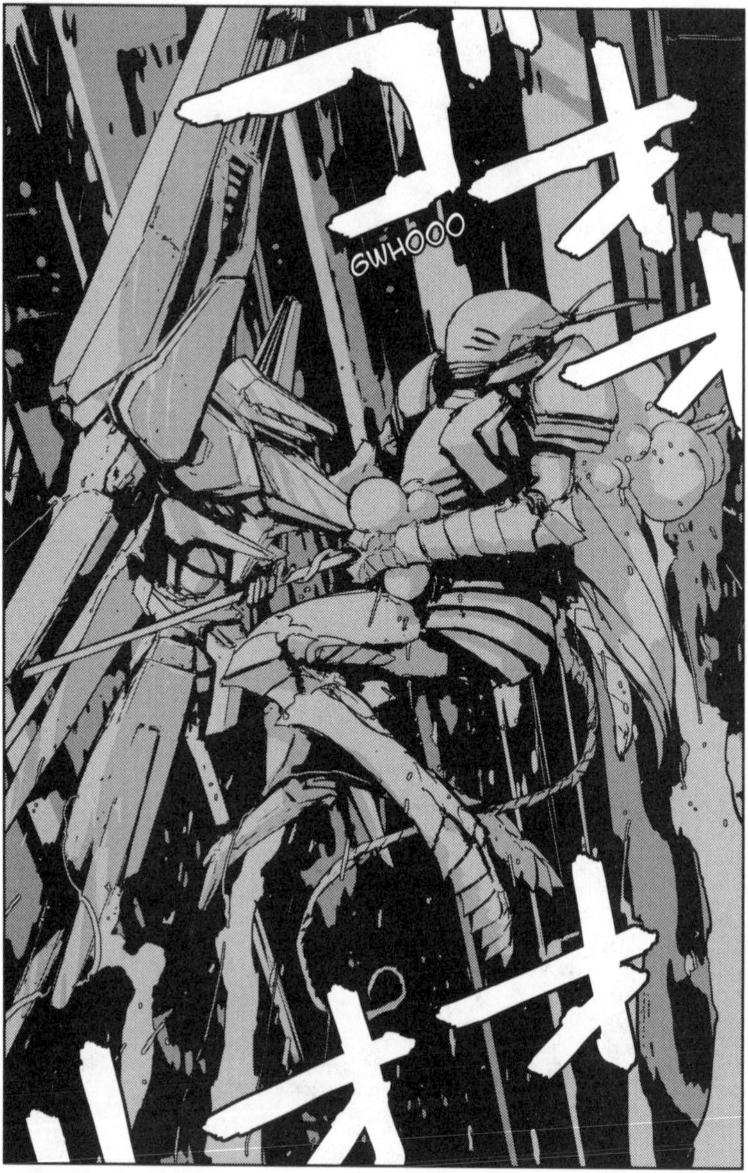
ゴオオオオ
GWHOOO

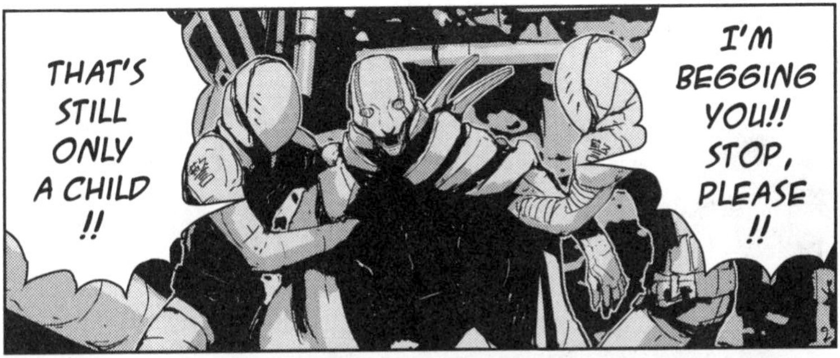
I'M BEGGING YOU!! STOP, PLEASE !!
THAT'S STILL ONLY A CHILD !!
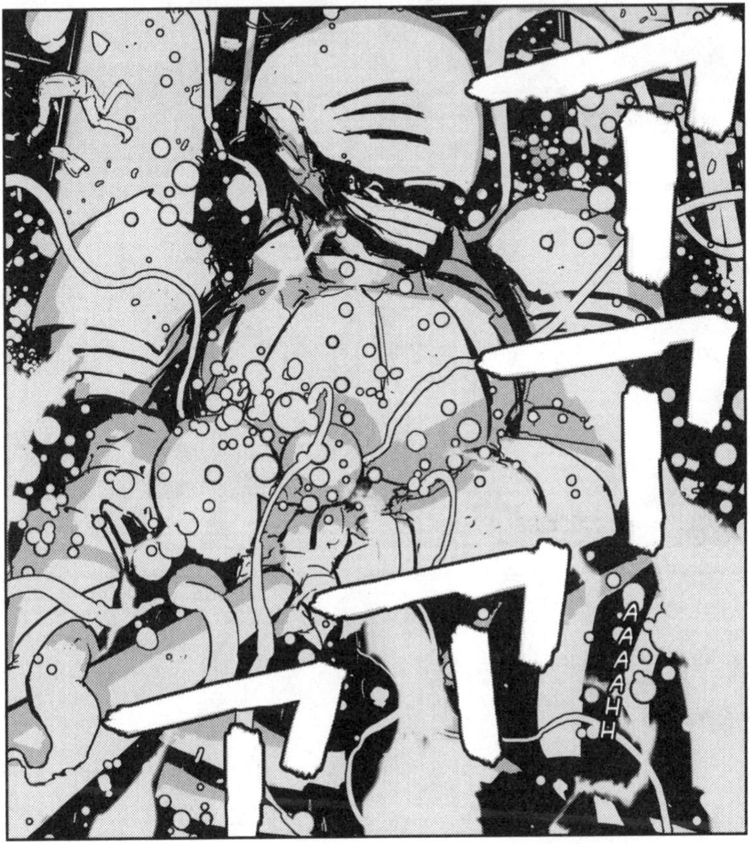
AAAAAHH

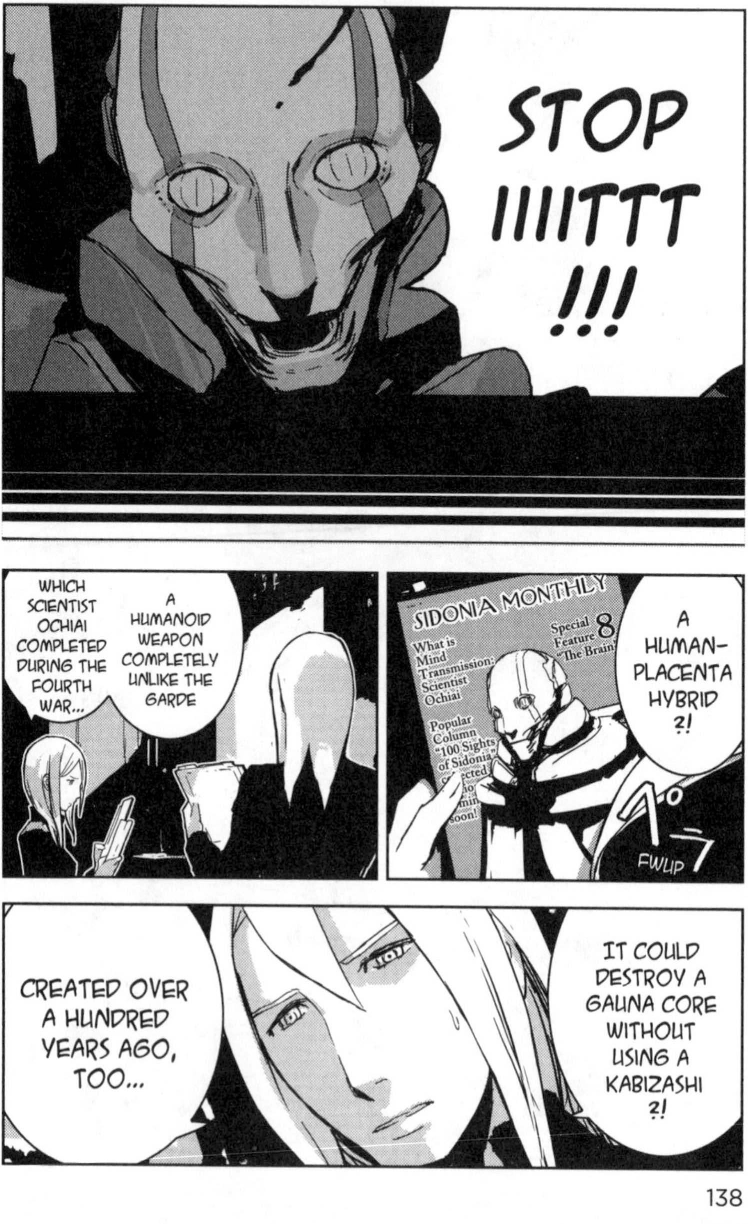
STOP IIIITTT !!!
SIDONIA MONTHLY
Special Feature 8 "The Brain"
What is Mind Transmission: Scientist Ochiai
Popular Column "100 Sights of Sidonia"
A HUMAN-PLACENTA HYBRID ?!
FWUP
A HUMANOID WEAPON COMPLETELY UNLIKE THE GARDE
WHICH SCIENTIST OCHIAI COMPLETED DURING THE FOURTH WAR...
IT COULD DESTROY A GAUNA CORE WITHOUT USING A KABIZASHI ?!
CREATED OVER A HUNDRED YEARS AGO, TOO...

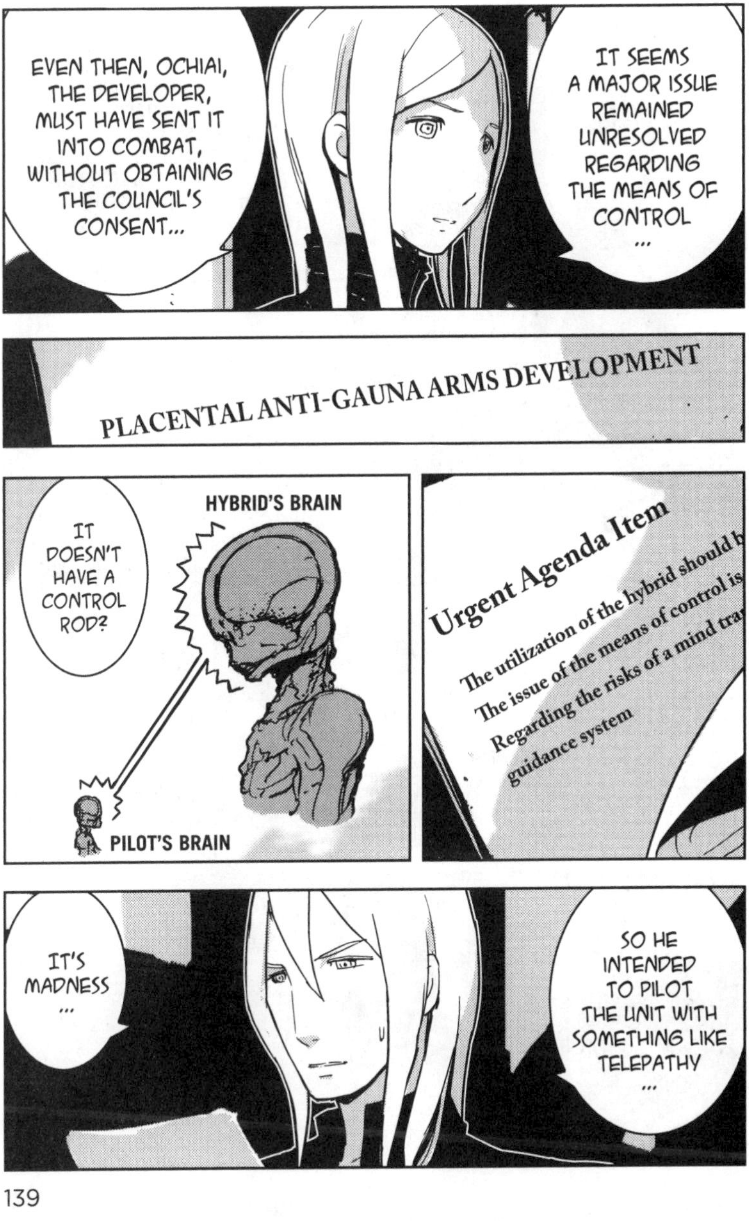
IT SEEMS A MAJOR ISSUE REMAINED UNRESOLVED REGARDING THE MEANS OF CONTROL ...
EVEN THEN, OCHIAI, THE DEVELOPER, MUST HAVE SENT IT INTO COMBAT, WITHOUT OBTAINING THE COUNCIL'S CONSENT...
PLACENTAL ANTI-GAUNA ARMS DEVELOPMENT
Urgent Agenda Item
The utilization of the hybrid should b
The issue of the means of control is
Regarding the risks of a mind tra
guidance system
HYBRID'S BRAIN
IT DOESN'T HAVE A CONTROL ROD?
PILOT'S BRAIN
SO HE INTENDED TO PILOT THE UNIT WITH SOMETHING LIKE TELEPATHY ...
IT'S MADNESS ...

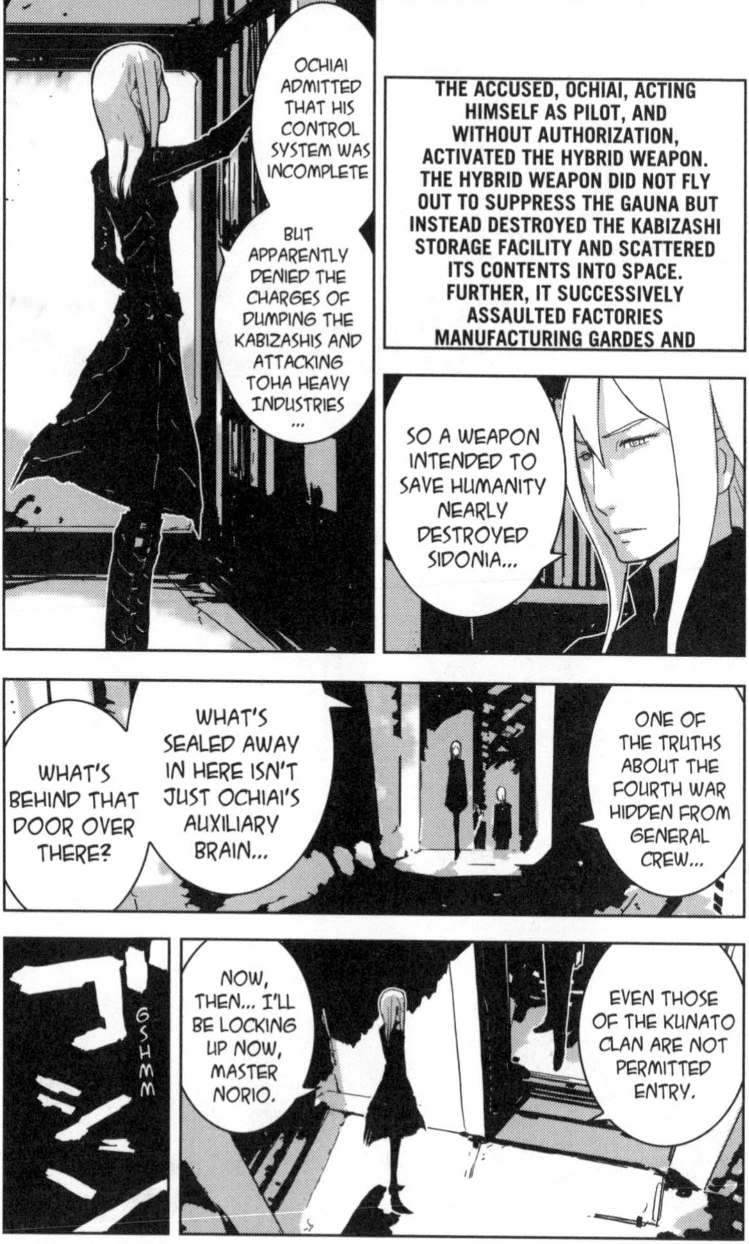
THE ACCUSED, OCHIAI, ACTING HIMSELF AS PILOT, AND WITHOUT AUTHORIZATION, ACTIVATED THE HYBRID WEAPON. THE HYBRID WEAPON DID NOT FLY OUT TO SUPPRESS THE GAUNA BUT INSTEAD DESTROYED THE KABIZASHI STORAGE FACILITY AND SCATTERED ITS CONTENTS INTO SPACE. FURTHER, IT SUCCESSIVELY ASSAULTED FACTORIES MANUFACTURING GARDES AND
SO A WEAPON INTENDED TO SAVE HUMANITY NEARLY DESTROYED SIDONIA...
OCHIAI ADMITTED THAT HIS CONTROL SYSTEM WAS INCOMPLETE
BUT APPARENTLY DENIED THE CHARGES OF DUMPING THE KABIZASHIS AND ATTACKING TOHA HEAVY INDUSTRIES ...
ONE OF THE TRUTHS ABOUT THE FOURTH WAR HIDDEN FROM GENERAL CREW...
WHAT'S SEALED AWAY IN HERE ISN'T JUST OCHIAI'S AUXILIARY BRAIN...
WHAT'S BEHIND THAT DOOR OVER THERE?
EVEN THOSE OF THE KUNATO CLAN ARE NOT PERMITTED ENTRY.
NOW, THEN... I'LL BE LOCKING UP NOW, MASTER NORIO.
GSHMM

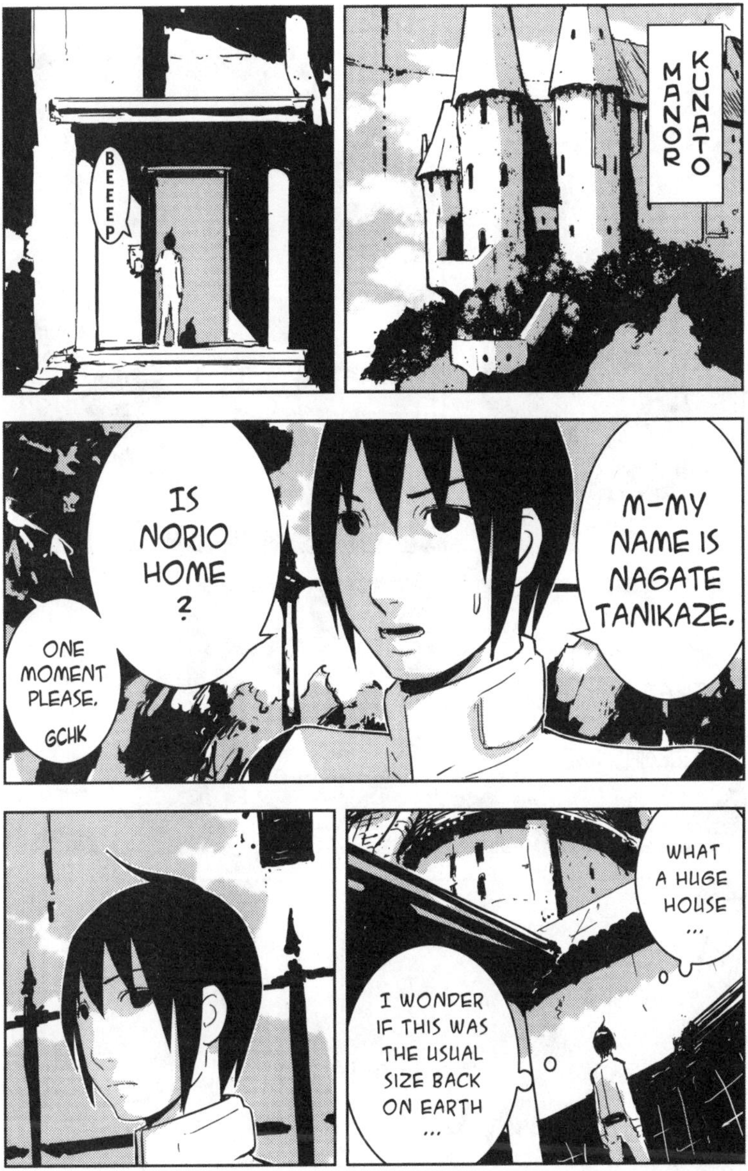
KUNATO MANOR
BEEEP
M-MY NAME IS NAGATE TANIKAZE.
IS NORIO HOME?
ONE MOMENT PLEASE.
GCHK
WHAT A HUGE HOUSE...
I WONDER IF THIS WAS THE USUAL SIZE BACK ON EARTH...

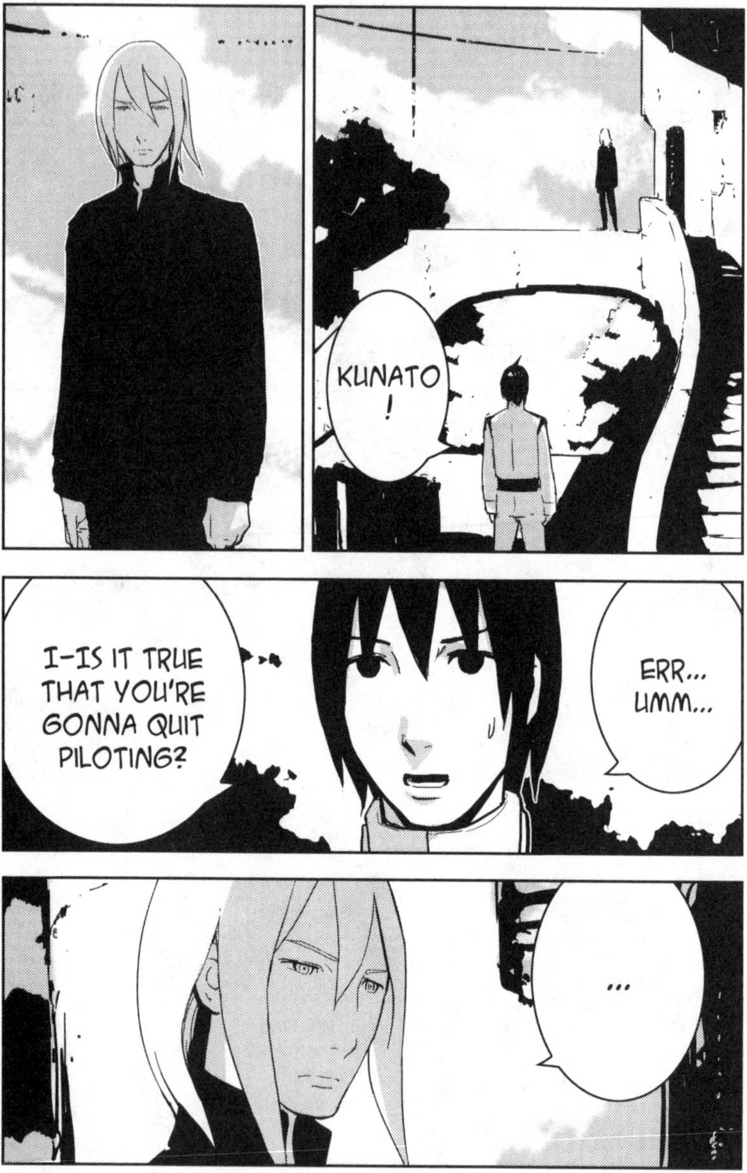
KUNATO!
ERR... UMM...
I-IS IT TRUE THAT YOU'RE GONNA QUIT PILOTING?
...

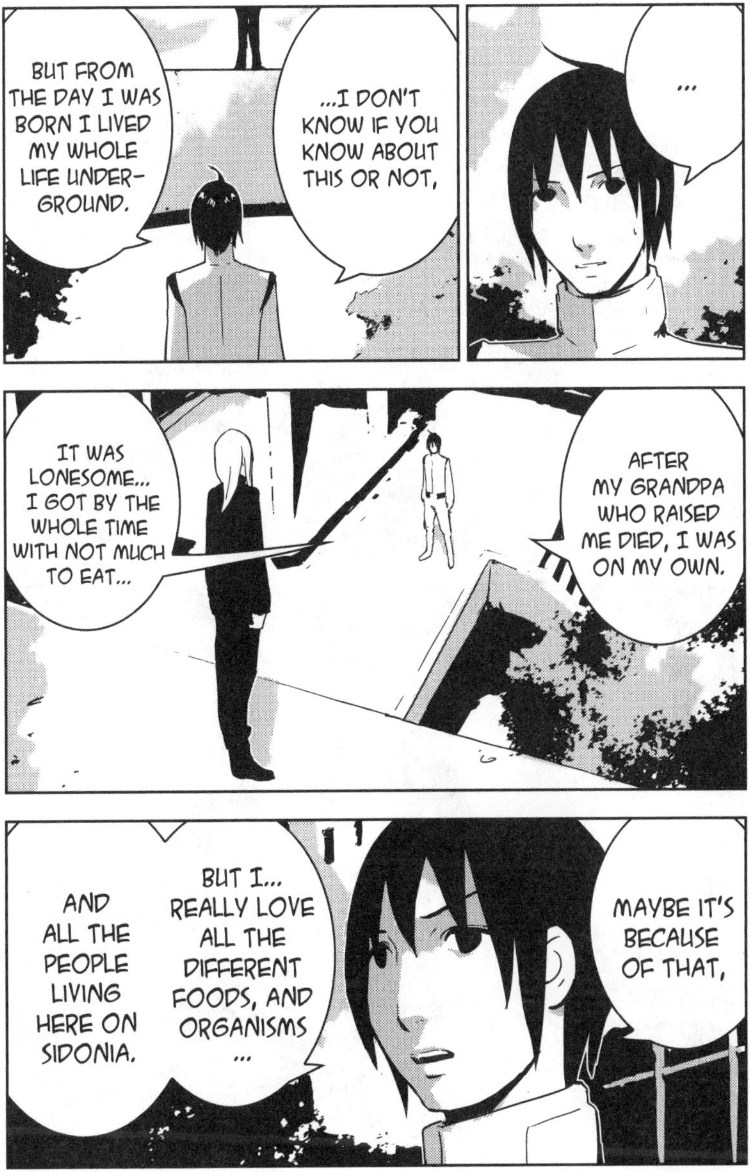
...
...I DON'T KNOW IF YOU KNOW ABOUT THIS OR NOT,
BUT FROM THE DAY I WAS BORN I LIVED MY WHOLE LIFE UNDER-GROUND.
AFTER MY GRANDPA WHO RAISED ME DIED, I WAS ON MY OWN.
IT WAS LONESOME... I GOT BY THE WHOLE TIME WITH NOT MUCH TO EAT...
MAYBE IT'S BECAUSE OF THAT,
BUT I... REALLY LOVE ALL THE DIFFERENT FOODS, AND ORGANISMS ...
AND ALL THE PEOPLE LIVING HERE ON SIDONIA.

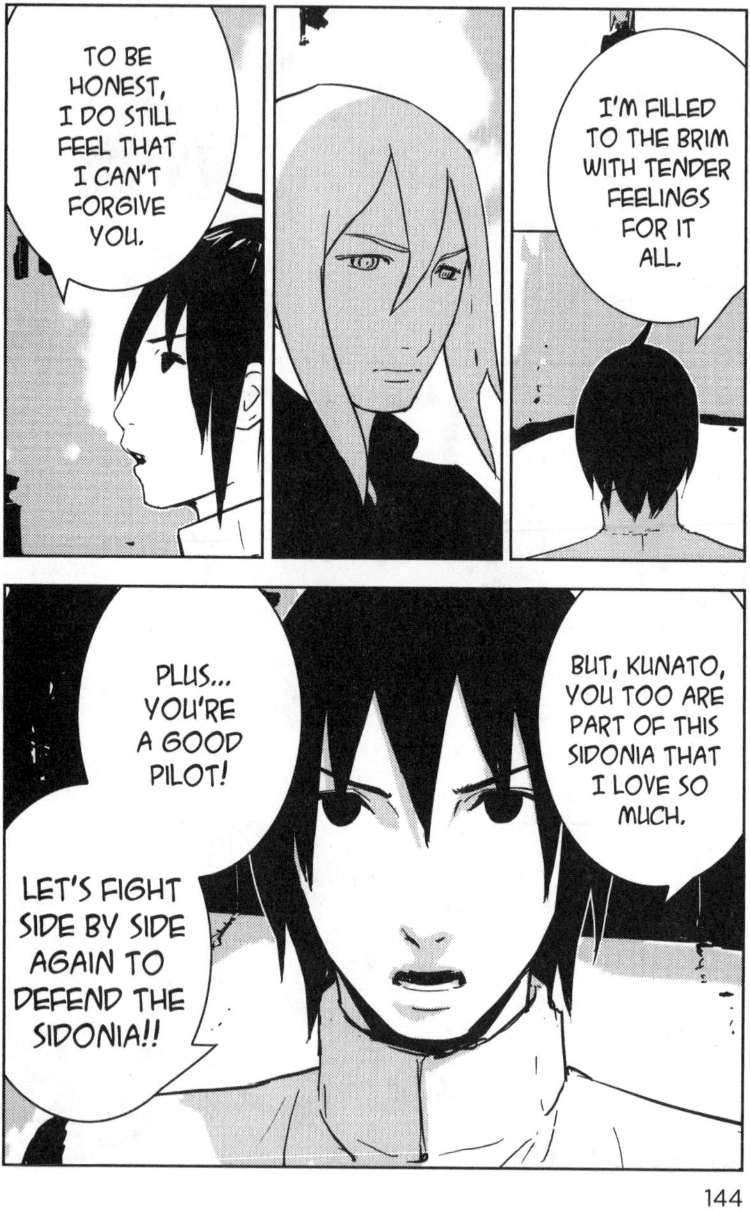
I'M FILLED TO THE BRIM WITH TENDER FEELINGS FOR IT ALL.
TO BE HONEST, I DO STILL FEEL THAT I CAN'T FORGIVE YOU.
BUT, KUNATO, YOU TOO ARE PART OF THIS SIDONIA THAT I LOVE SO MUCH.
PLUS... YOU'RE A GOOD PILOT!
LET'S FIGHT SIDE BY SIDE AGAIN TO DEFEND THE SIDONIA!!

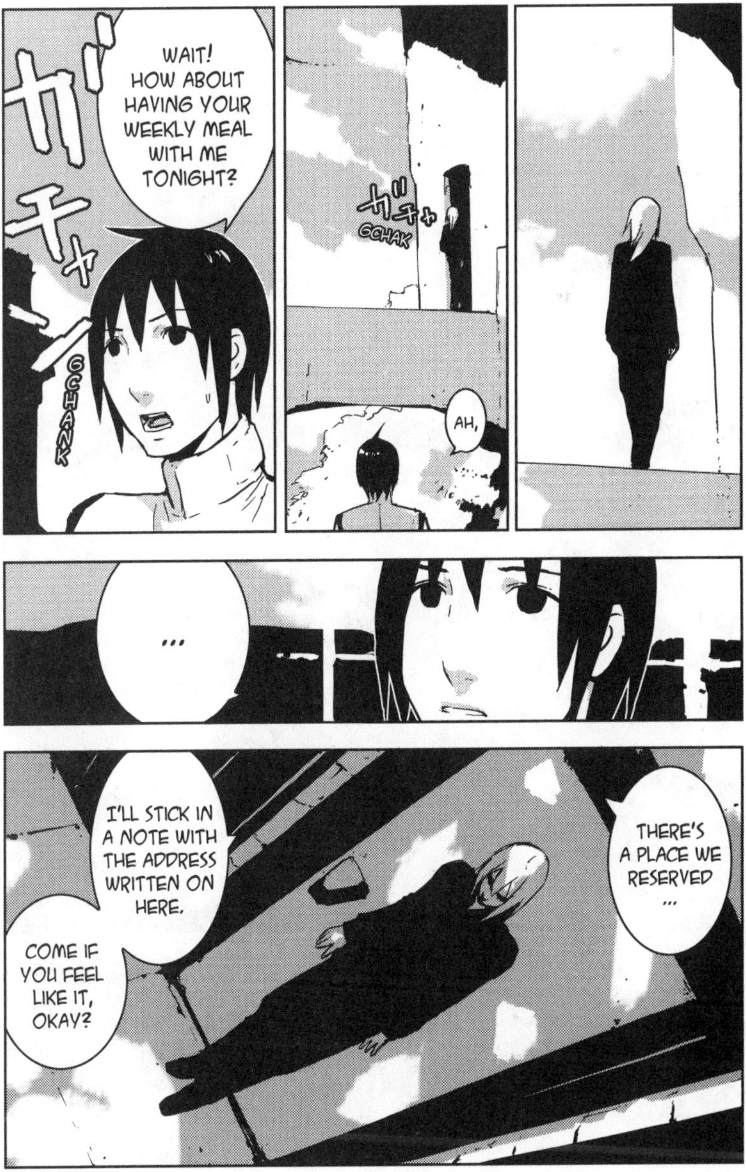
GCHAK
AH,
WAIT! HOW ABOUT HAVING YOUR WEEKLY MEAL WITH ME TONIGHT?
GCHANK
...
THERE'S A PLACE WE RESERVED ...
I'LL STICK IN A NOTE WITH THE ADDRESS WRITTEN ON HERE.
COME IF YOU FEEL LIKE IT, OKAY?

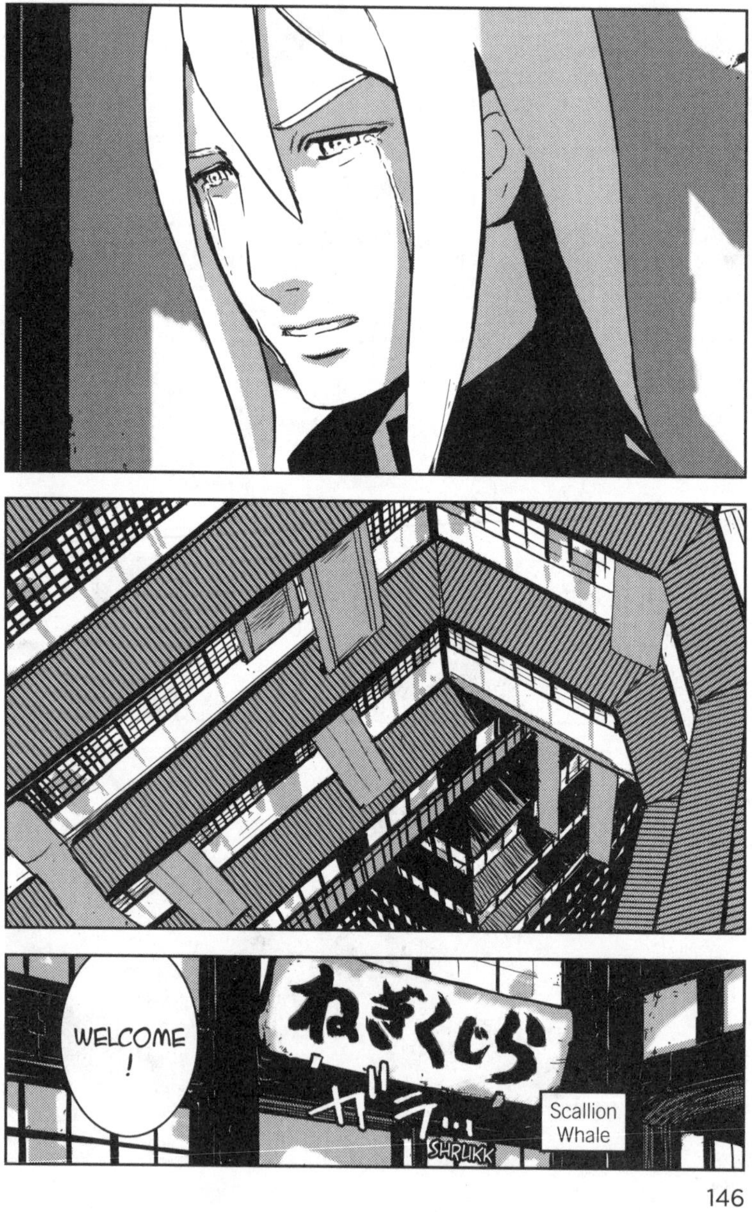
ねぎくじら
Scallion Whale
ガラ…
SHRUKK
WELCOME!

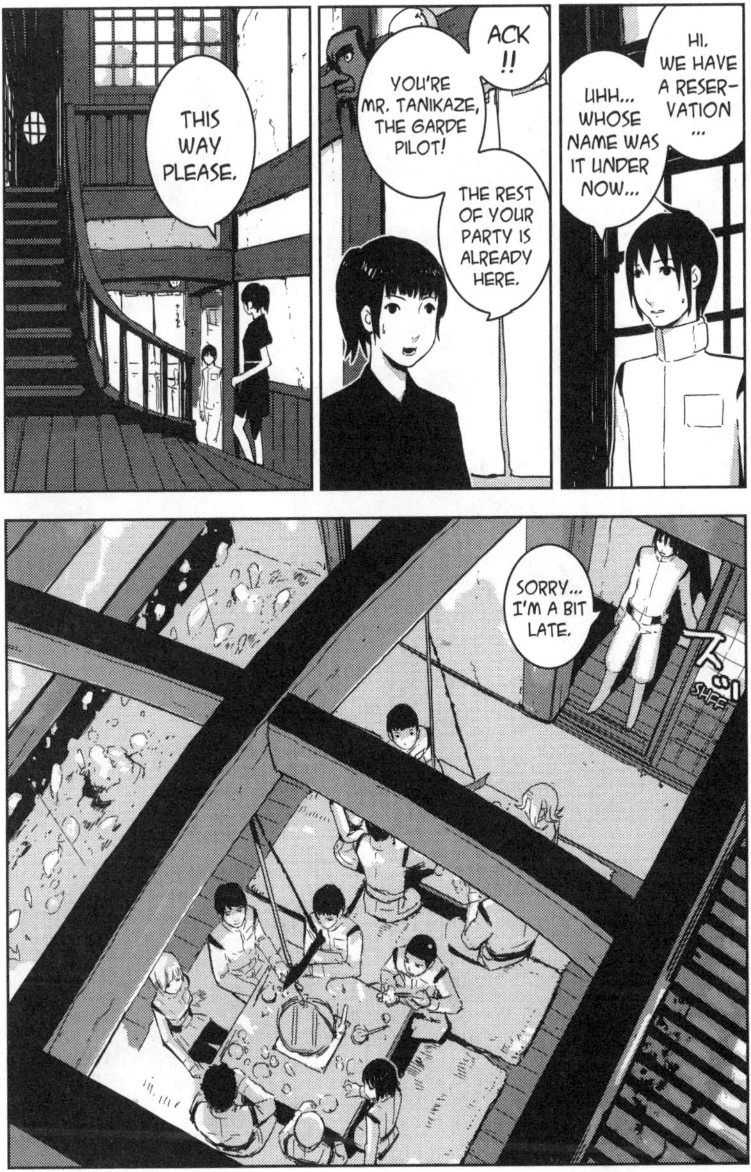
HI. WE HAVE A RESER-VATION ...
UHH... WHOSE NAME WAS IT UNDER NOW...
ACK !!
YOU'RE MR. TANIKAZE, THE GARDE PILOT!
THE REST OF YOUR PARTY IS ALREADY HERE.
THIS WAY PLEASE.
SORRY... I'M A BIT LATE.
ス
SHFF

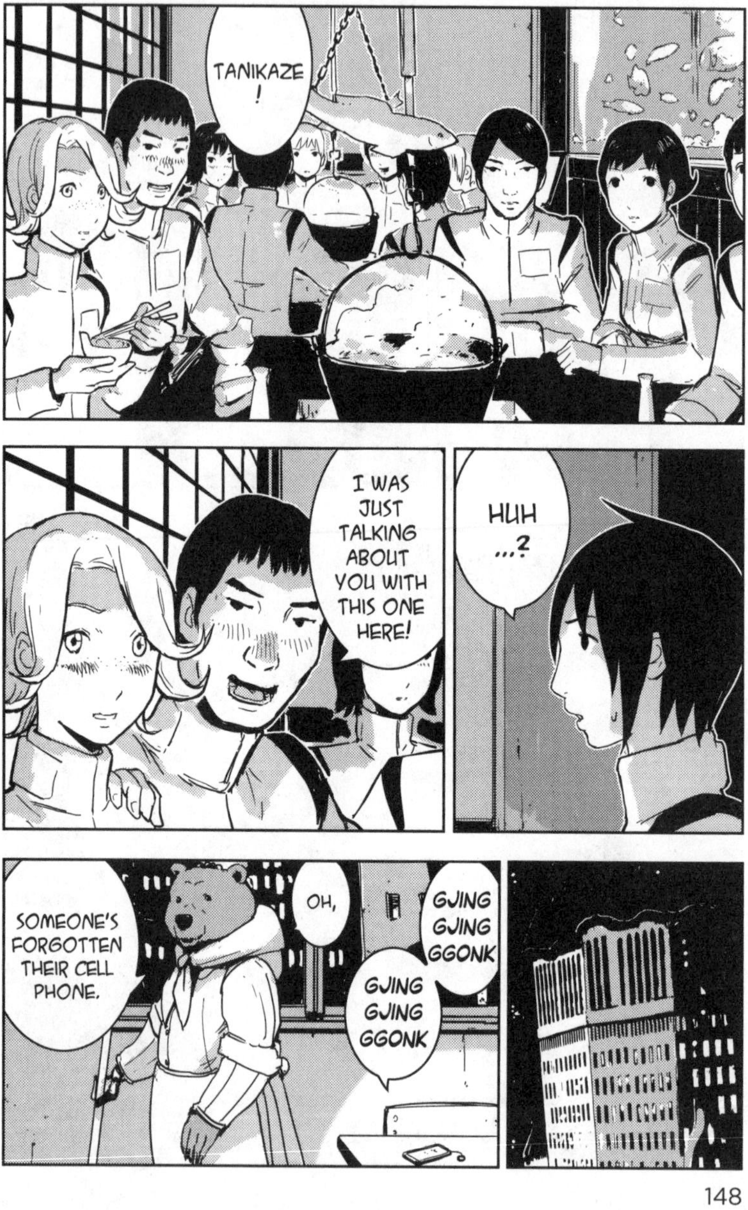
TANIKAZE !
HUH ...?
I WAS JUST TALKING ABOUT YOU WITH THIS ONE HERE!
GJING GJING GGONK
OH,
GJING GJING GGONK
SOMEONE'S FORGOTTEN THEIR CELL PHONE.

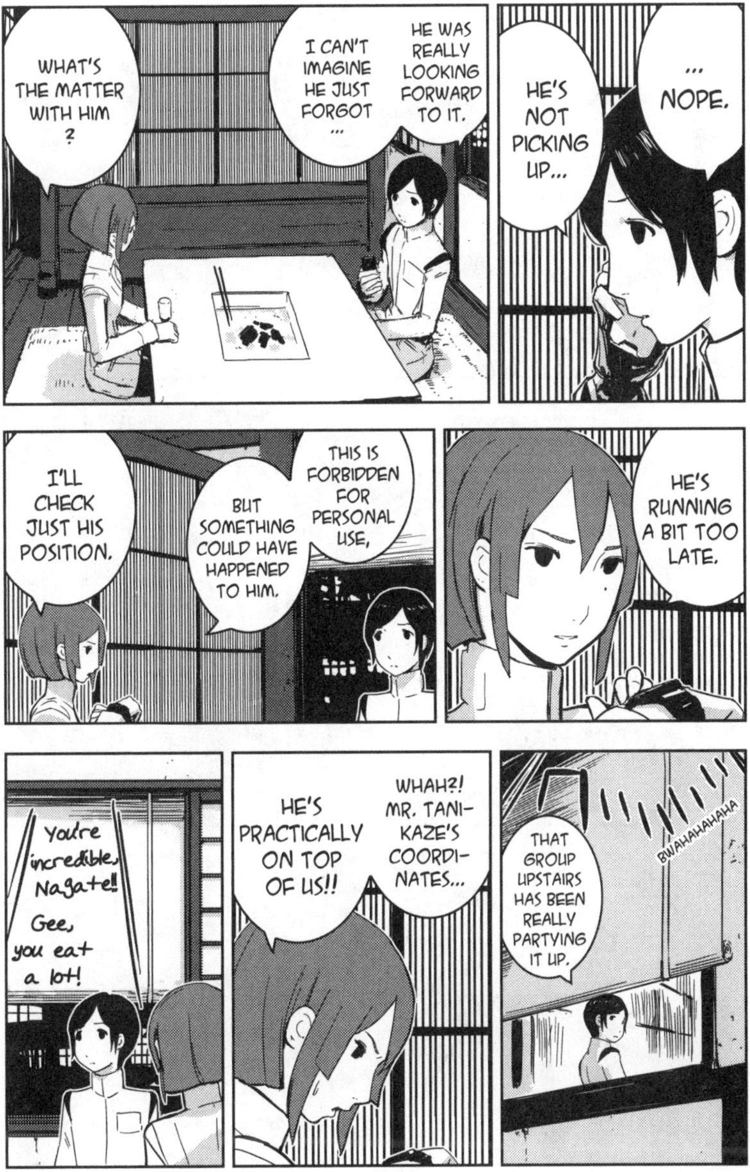
... NOPE.
HE'S NOT PICKING UP...
HE WAS REALLY LOOKING FORWARD TO IT.
I CAN'T IMAGINE HE JUST FORGOT ...
WHAT'S THE MATTER WITH HIM ?
HE'S RUNNING A BIT TOO LATE.
THIS IS FORBIDDEN FOR PERSONAL USE,
BUT SOMETHING COULD HAVE HAPPENED TO HIM.
I'LL CHECK JUST HIS POSITION.
BWAHAHAHAHA
THAT GROUP UPSTAIRS HAS BEEN REALLY PARTYING IT UP.
WHAH?! MR. TANI-KAZE'S COORDI-NATES...
HE'S PRACTICALLY ON TOP OF US!!
You're incredible, Nagate!!
Gee, you eat a lot!

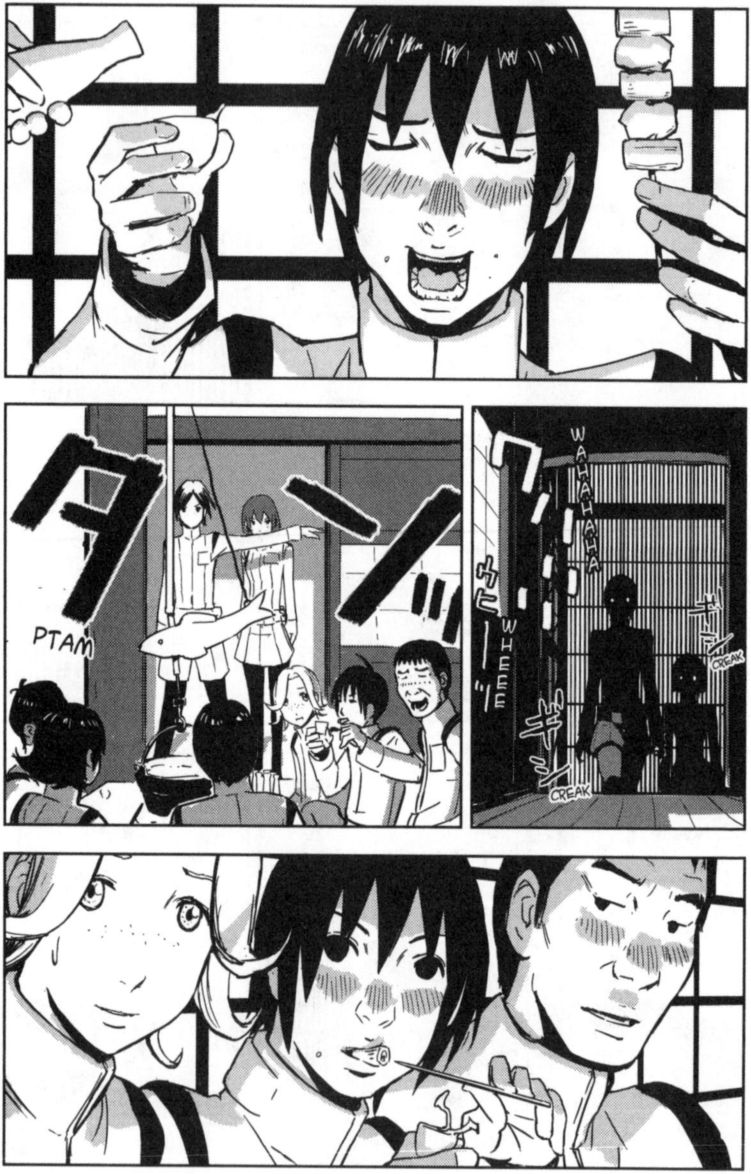
WAHAHAHA
WHEEE
CREAK
CREAK
PTAM

NAGATE
!!

WHLIDD

KRIK

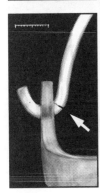

GDNK

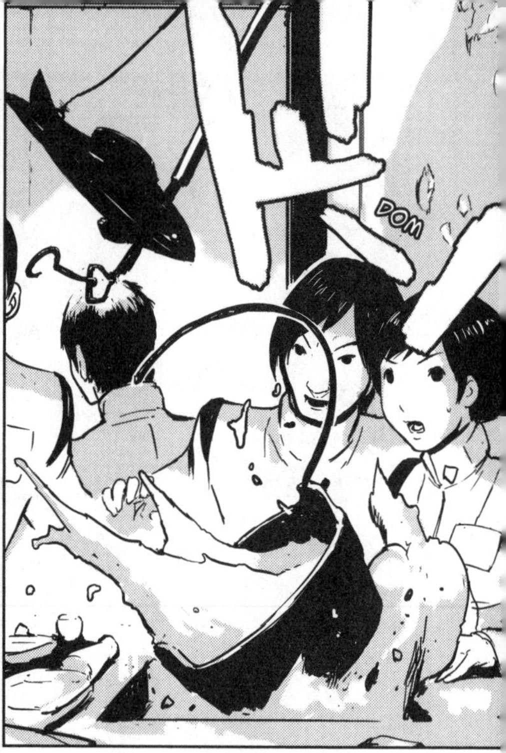
DOM
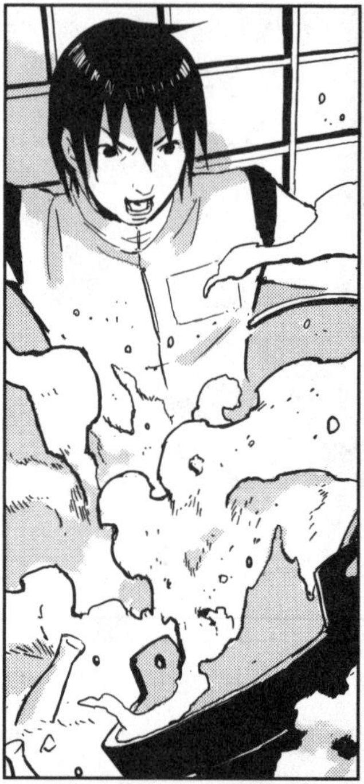
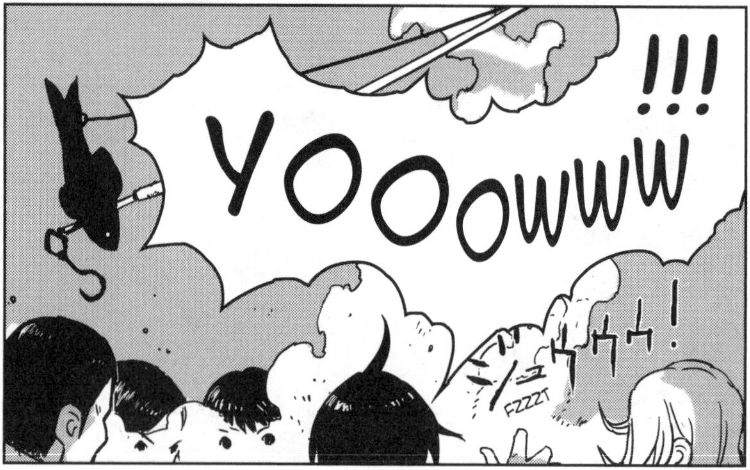
YOOOWWW!!!
FZZZT

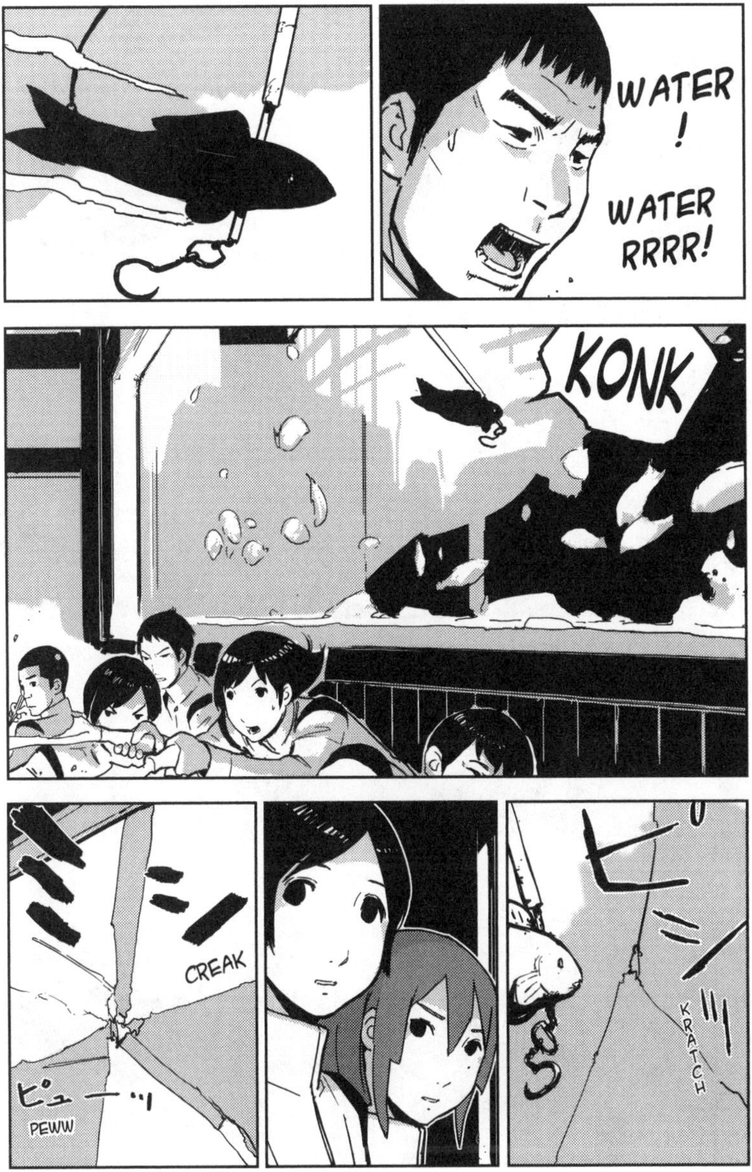
WATER!
WATER RRRR!
KONK
KRATCH
CREAK
ピューーッ
PEWW

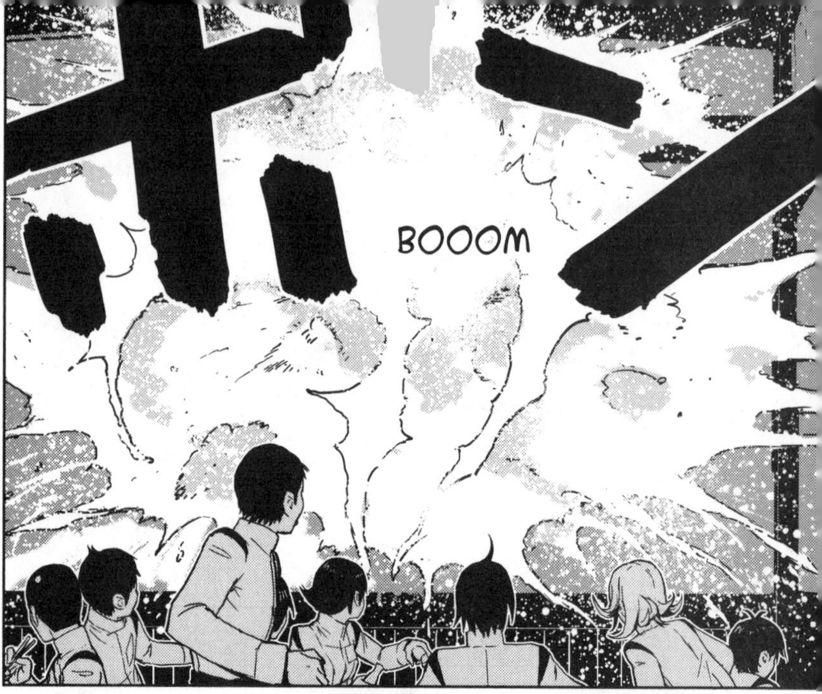
BOOOM
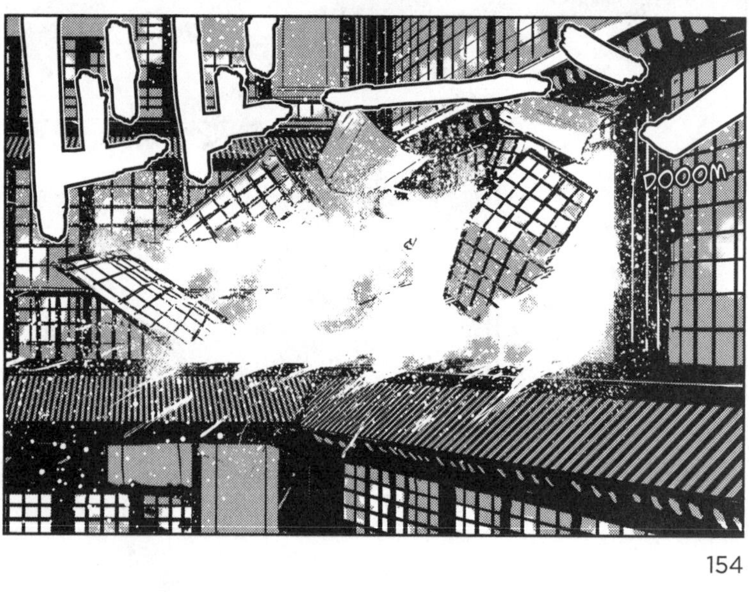
DOOOM

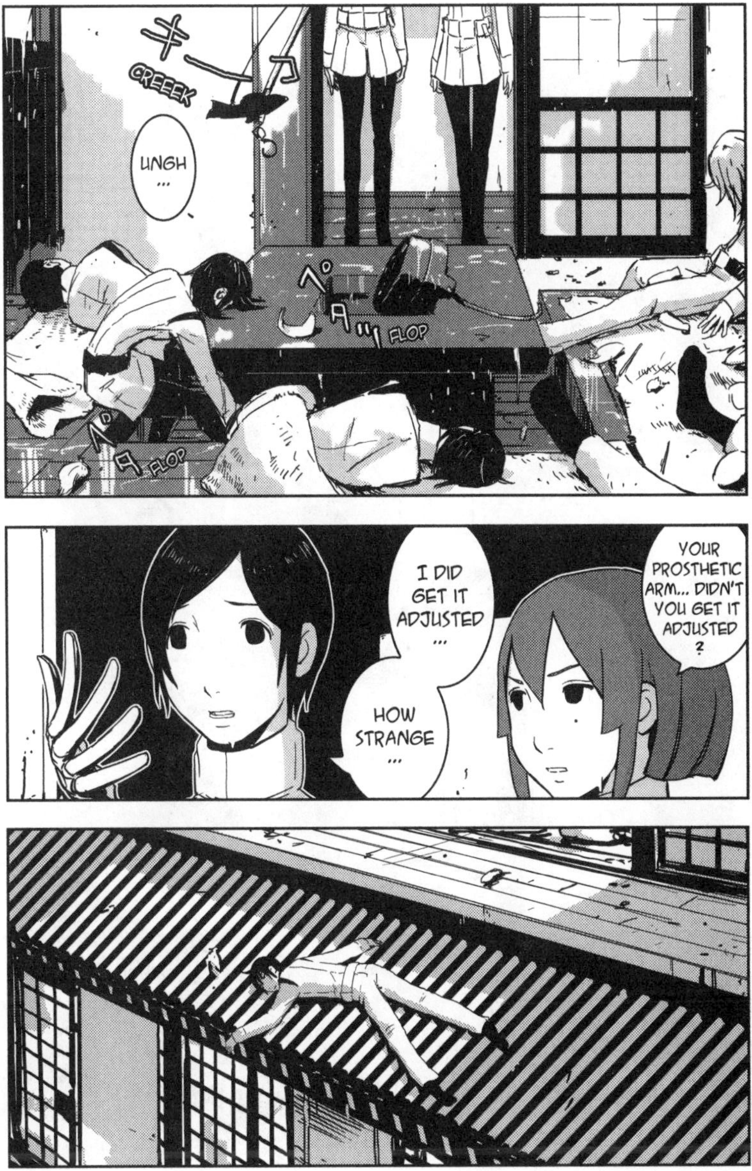
CREEEK
UNGH...
FLOP
FLOP
YOUR PROSTHETIC ARM... DIDN'T YOU GET IT ADJUSTED?
I DID GET IT ADJUSTED...
HOW STRANGE...

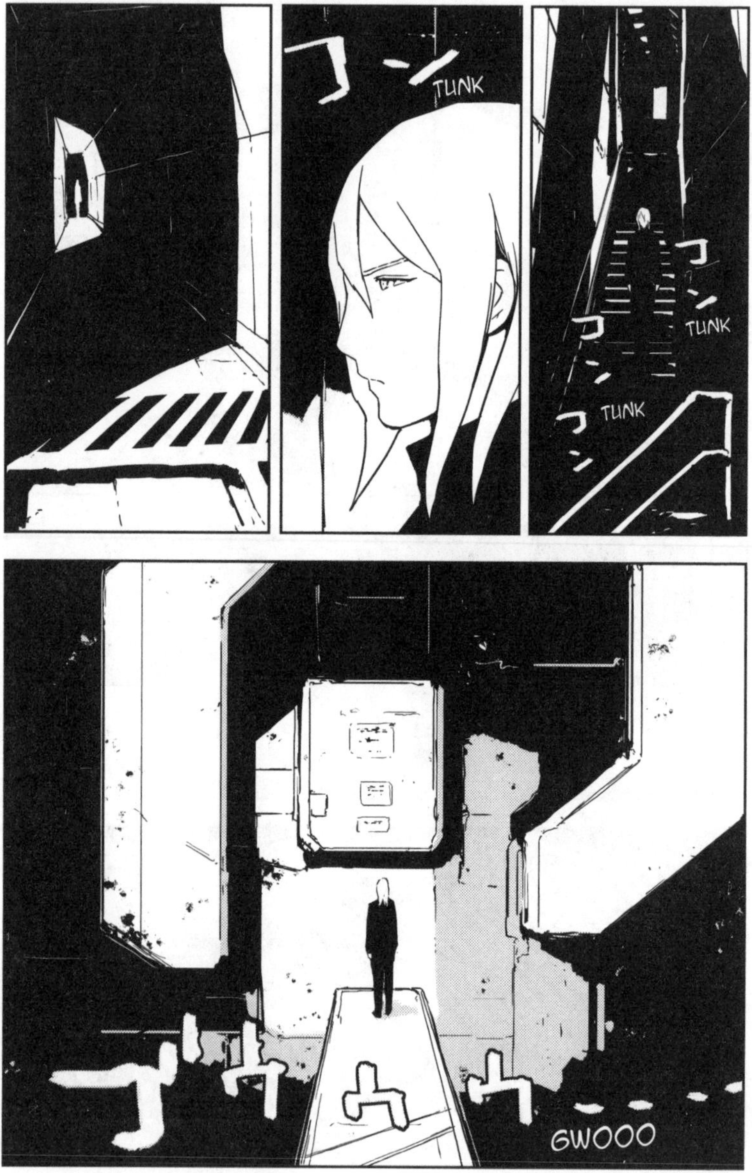
コン
TUNK
コン
コン
TUNK
コーン
TUNK
ブゥゥゥ……
GWOOO

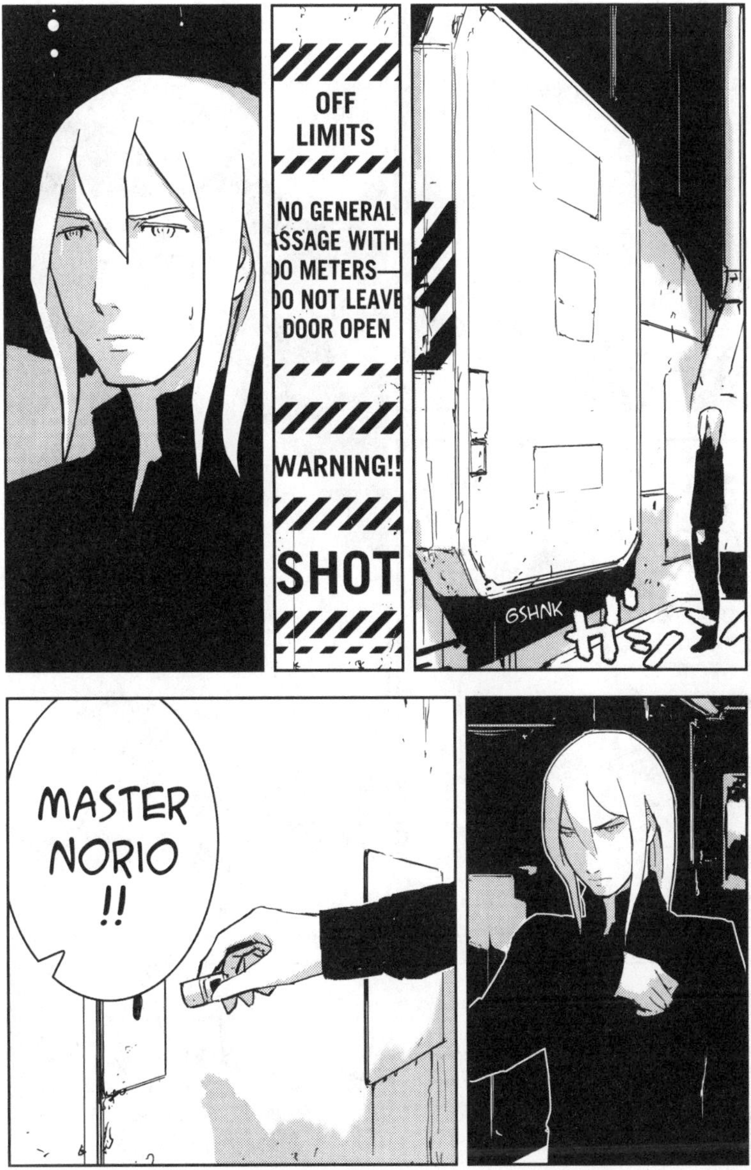
GSHNK
OFF LIMITS
NO GENERAL
ASSAGE WITH
O METERS—
DO NOT LEAVE
DOOR OPEN
WARNING!!
SHOT
MASTER NORIO !!

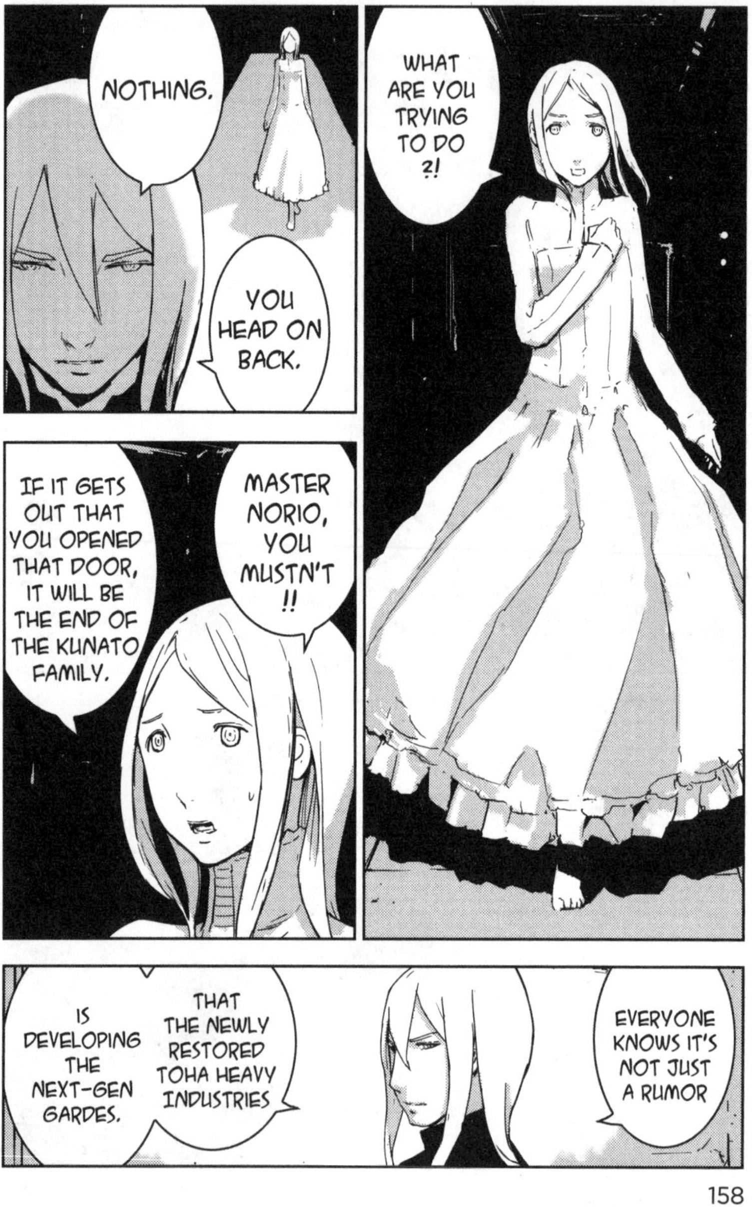
WHAT ARE YOU TRYING TO DO ?!
NOTHING.
YOU HEAD ON BACK.
MASTER NORIO, YOU MUSTN'T !!
IF IT GETS OUT THAT YOU OPENED THAT DOOR, IT WILL BE THE END OF THE KUNATO FAMILY.
EVERYONE KNOWS IT'S NOT JUST A RUMOR
THAT THE NEWLY RESTORED TOHA HEAVY INDUSTRIES
IS DEVELOPING THE NEXT-GEN GARDES.

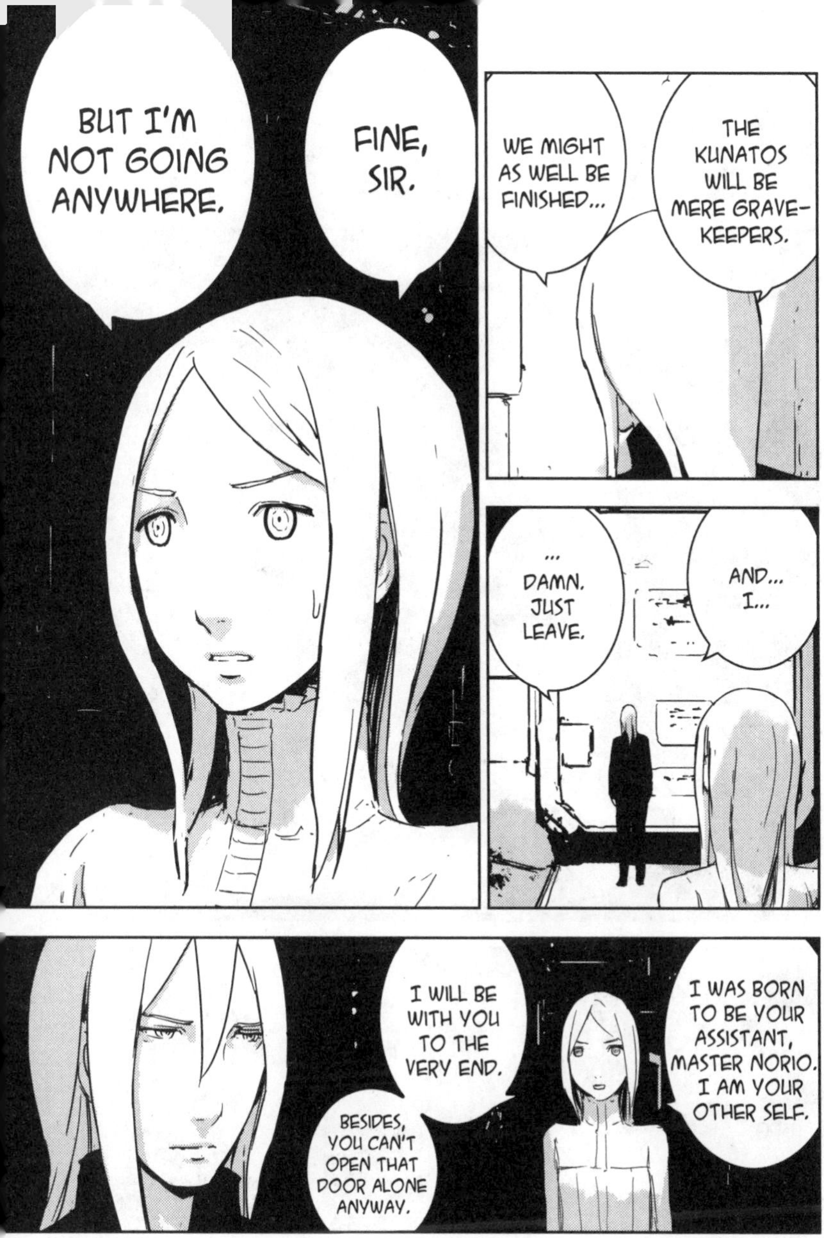
THE KUNATOS WILL BE MERE GRAVE-KEEPERS.
WE MIGHT AS WELL BE FINISHED...
AND... I...
... DAMN. JUST LEAVE.
FINE, SIR.
BUT I'M NOT GOING ANYWHERE.
I WAS BORN TO BE YOUR ASSISTANT, MASTER NORIO. I AM YOUR OTHER SELF.
I WILL BE WITH YOU TO THE VERY END.
BESIDES, YOU CAN'T OPEN THAT DOOR ALONE ANYWAY.

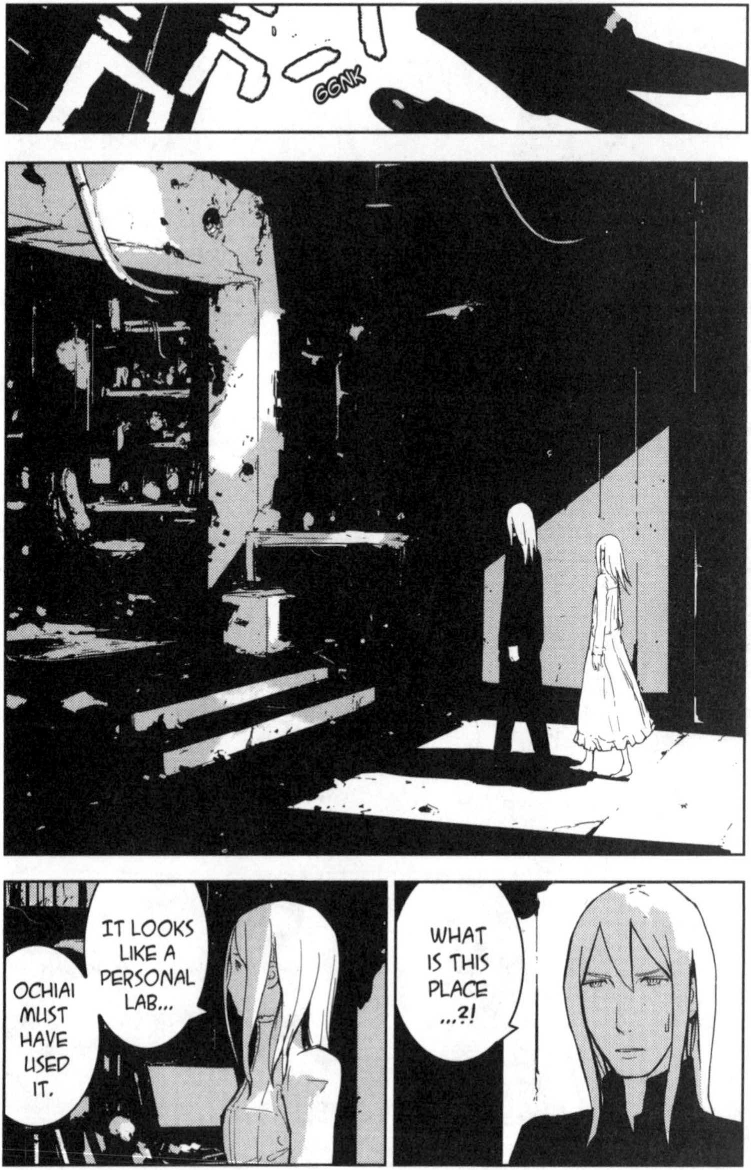
GGNK
WHAT IS THIS PLACE ...?!
IT LOOKS LIKE A PERSONAL LAB...
OCHIAI MUST HAVE USED IT.

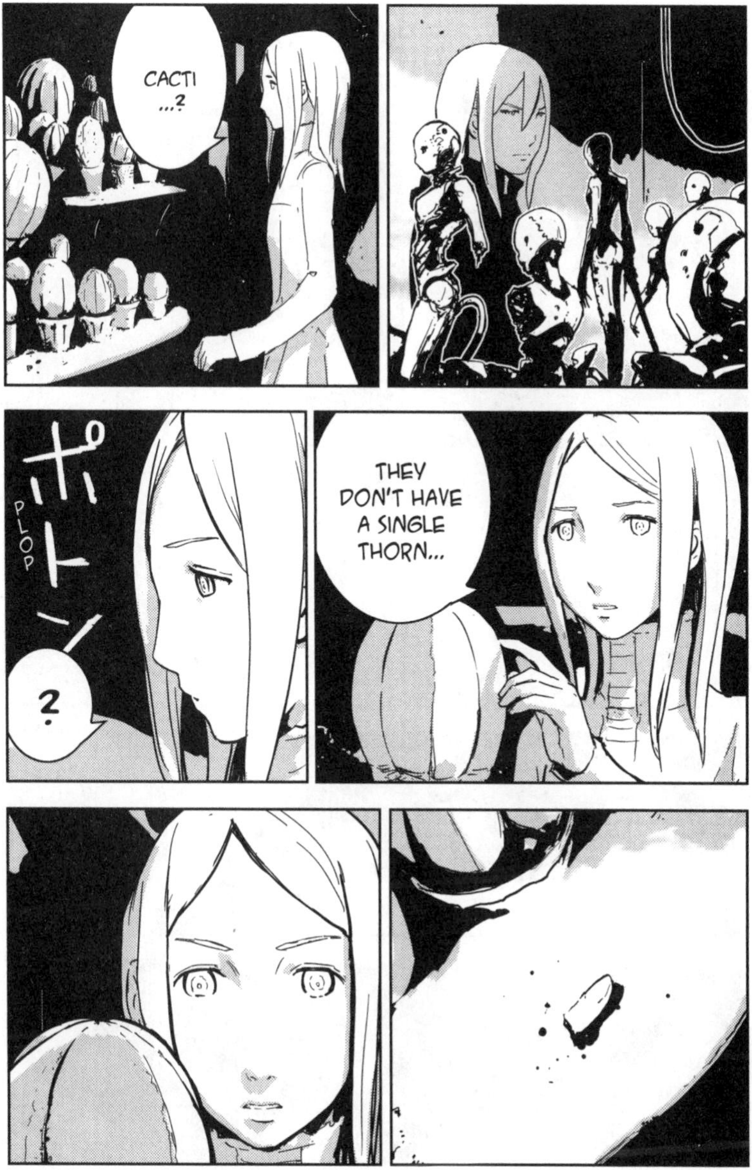
CACTI ...?
THEY DON'T HAVE A SINGLE THORN...
PLOP
?

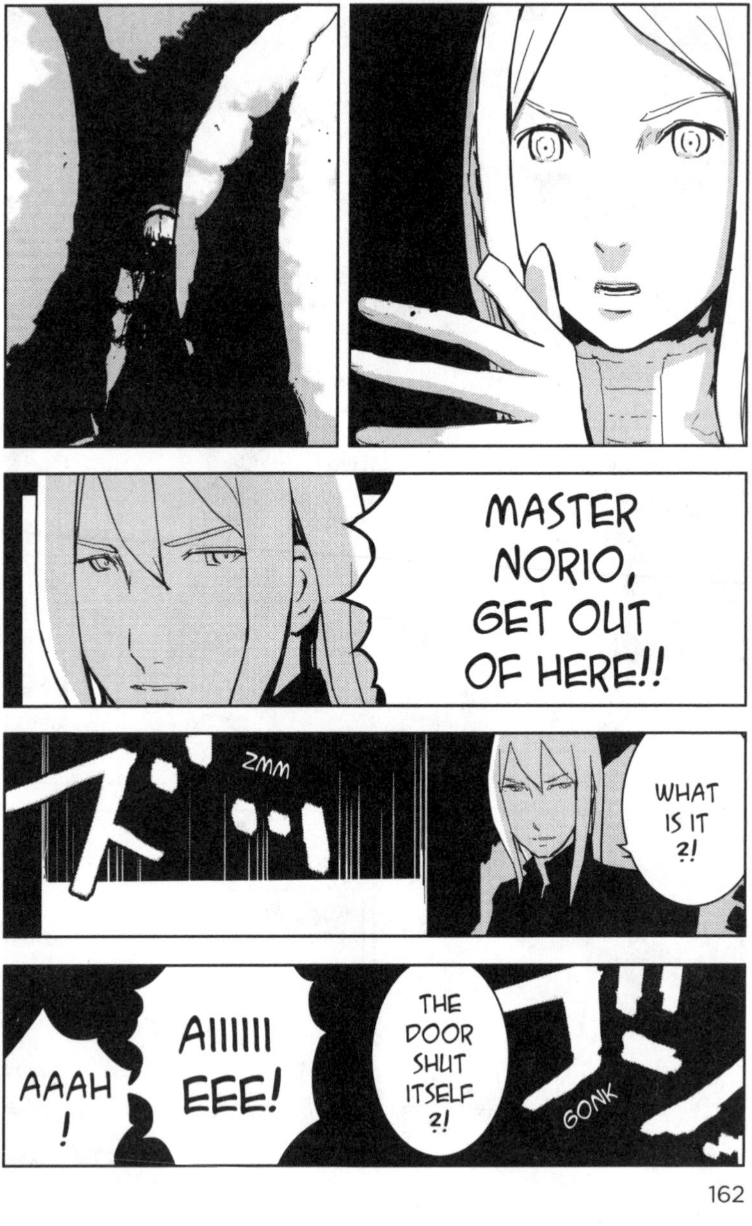
MASTER NORIO, GET OUT OF HERE!!
WHAT IS IT ?!
ZMM
GONK
THE DOOR SHUT ITSELF ?!
AIIIIII EEE!
AAAH !

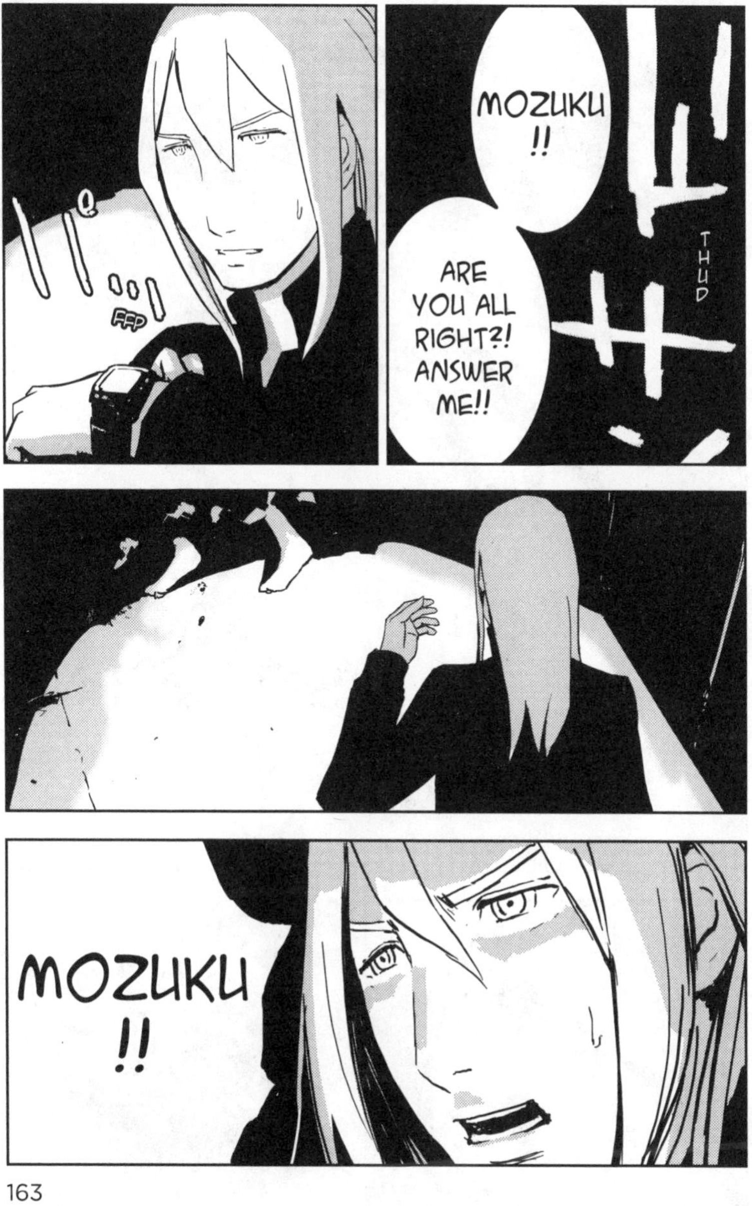
MOZUKU !!
ARE YOU ALL RIGHT?! ANSWER ME!!
THUD
FFP
MOZUKU !!

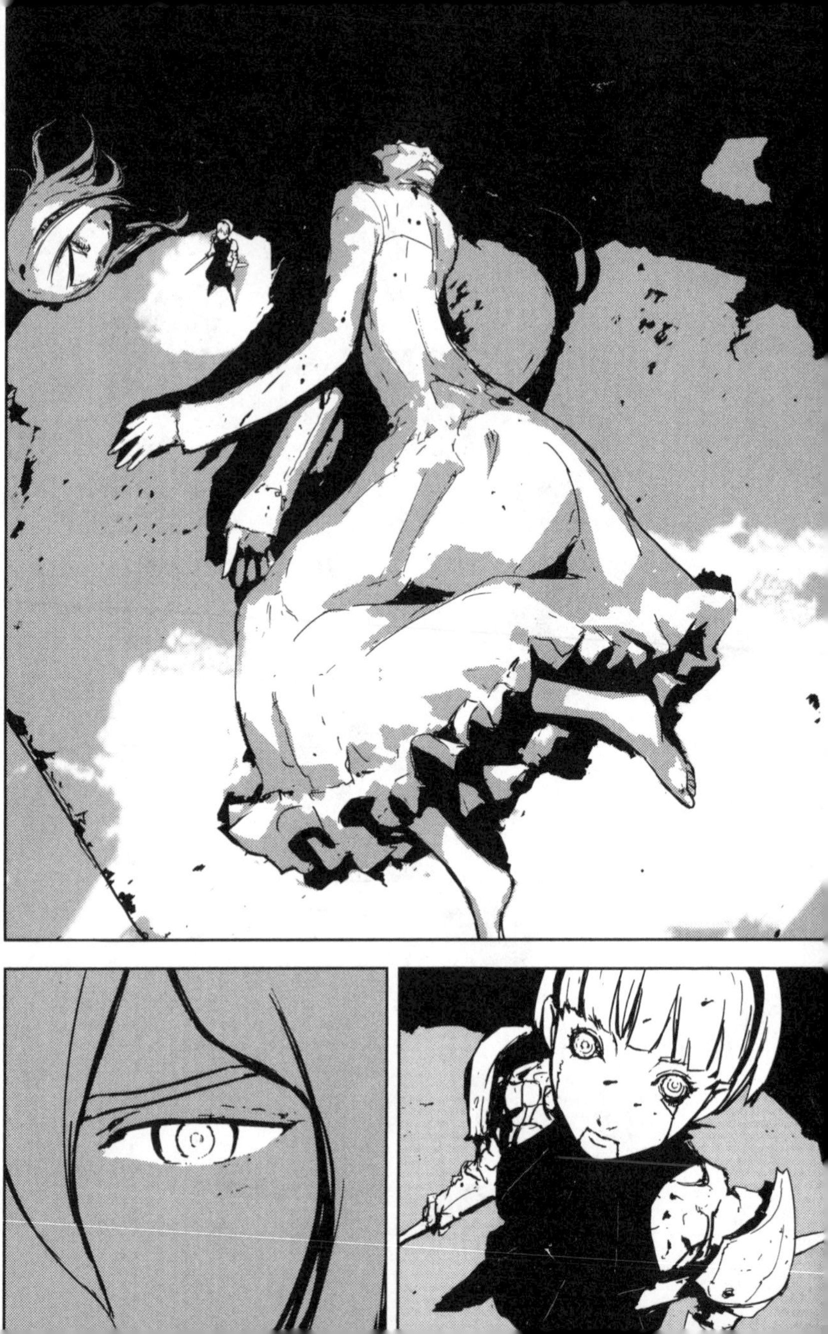

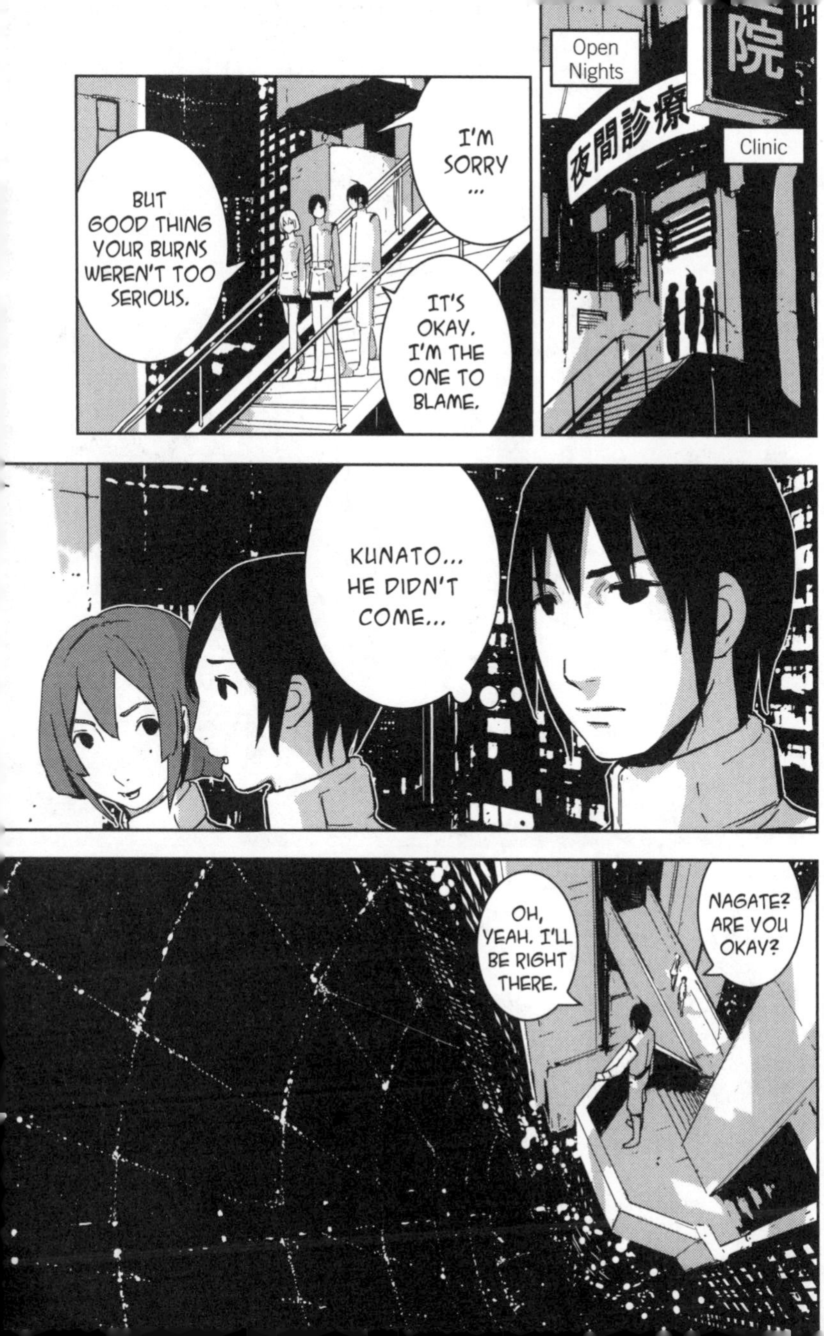
Open Nights
院
夜間診療
Clinic
I'M SORRY ...
BUT GOOD THING YOUR BURNS WEREN'T TOO SERIOUS.
IT'S OKAY. I'M THE ONE TO BLAME.
KUNATO... HE DIDN'T COME...
NAGATE? ARE YOU OKAY?
OH, YEAH. I'LL BE RIGHT THERE.

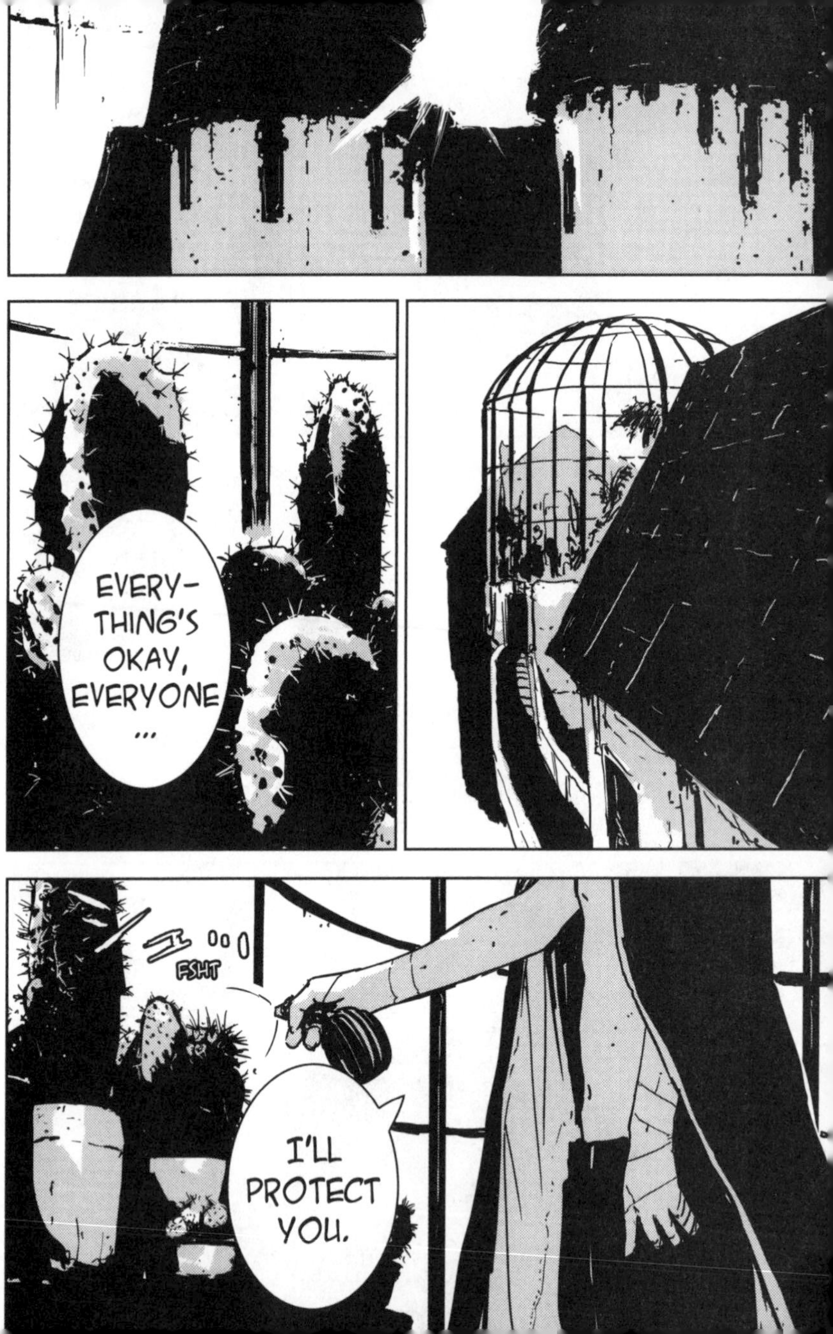
EVERY-THING'S OKAY, EVERYONE ...
FSHT
I'LL PROTECT YOU.

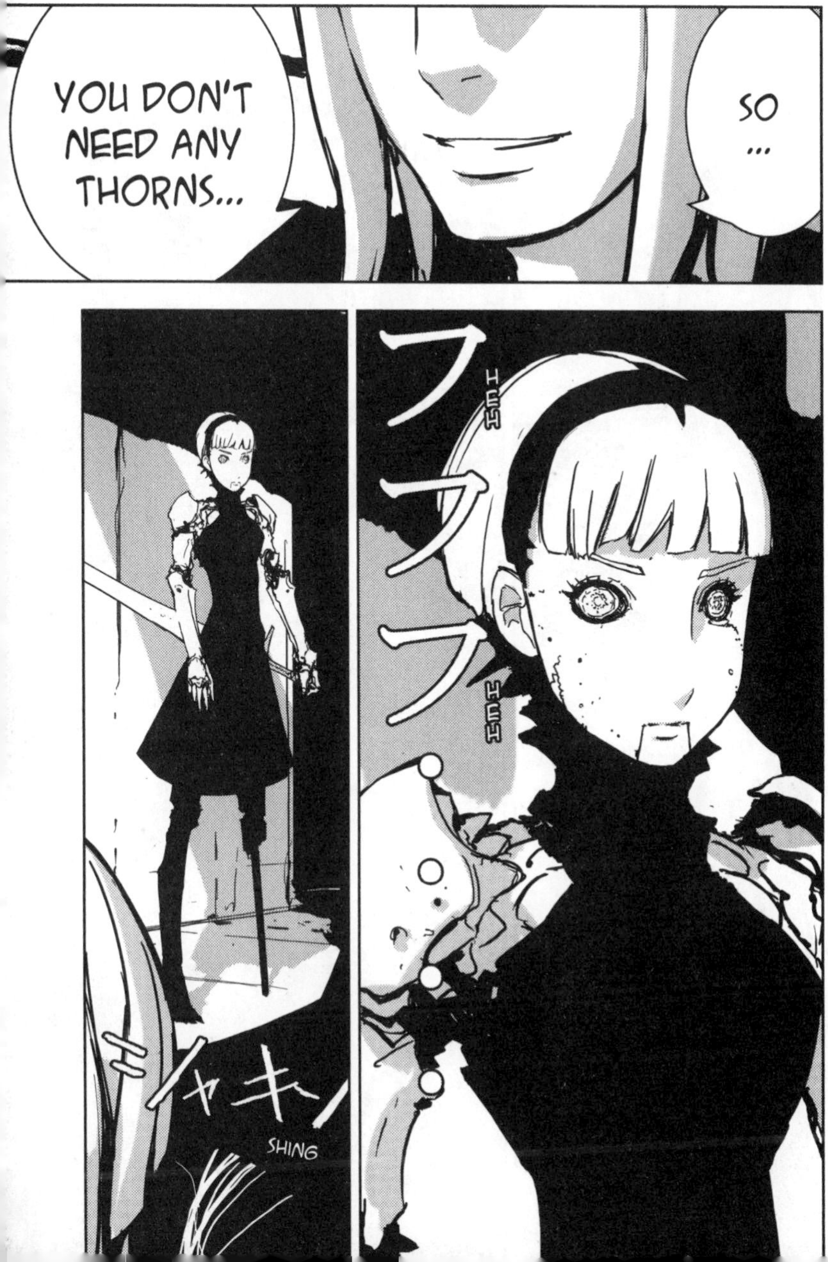
SO ...
YOU DON'T NEED ANY THORNS...
フフフ
HEH HEH
ジャキ
SHING

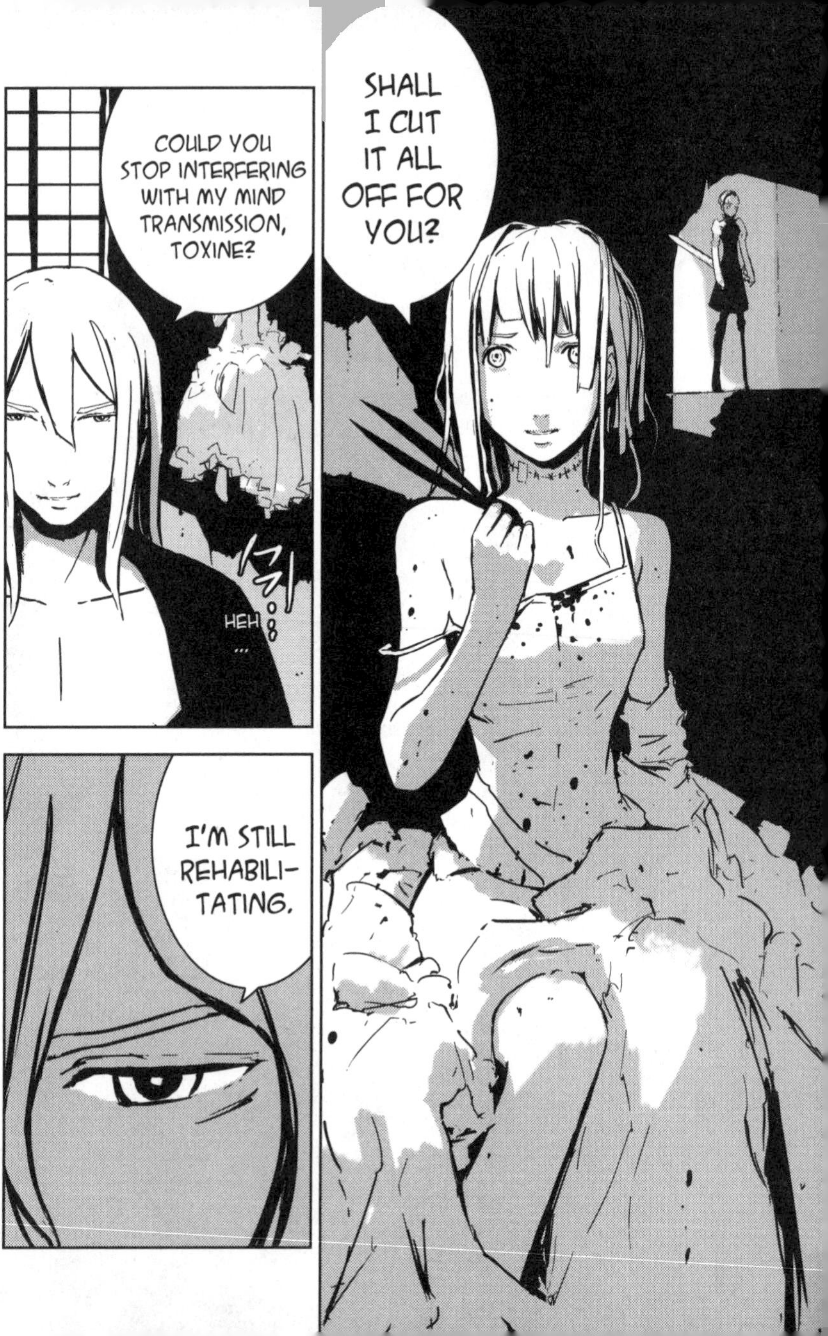
SHALL I CUT IT ALL OFF FOR YOU?
COULD YOU STOP INTERFERING WITH MY MIND TRANSMISSION, TOXINE?
HEH...
I'M STILL REHABILI-TATING.

KNIGHTS OF SIDONIA Volume 4: END

附録
SUPPLEMENT
KNIGHTS OF SIDONIA

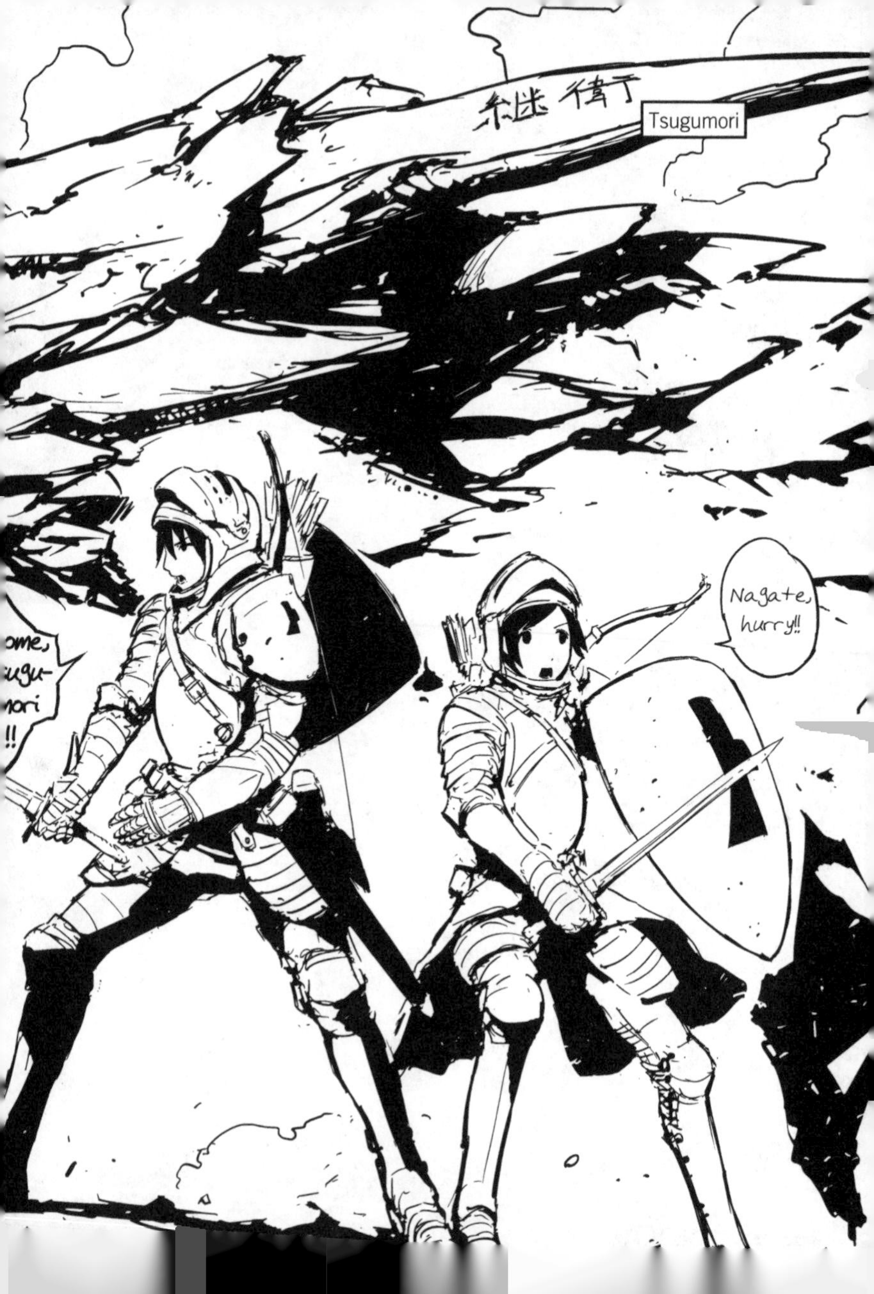
継衛
Tsugumori
Nagate, hurry!!
ome, ugu- nori !!

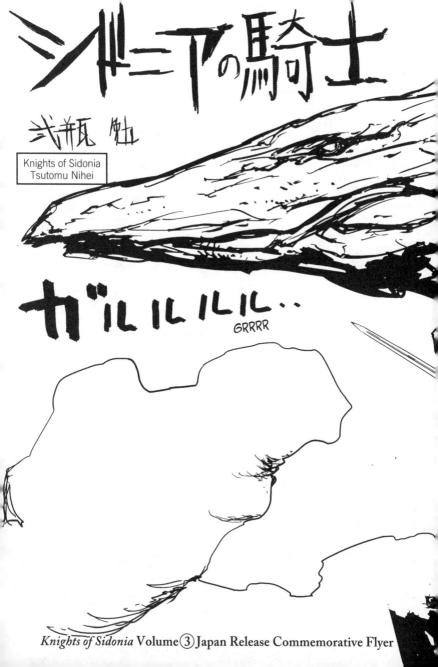

シドニアの騎士

弐瓶 勉

Knights of Sidonia
Tsutomu Nihei

ガ゛ルルルル‥
GRRRR

Knights of Sidonia Volume ③ Japan Release Commemorative Flyer

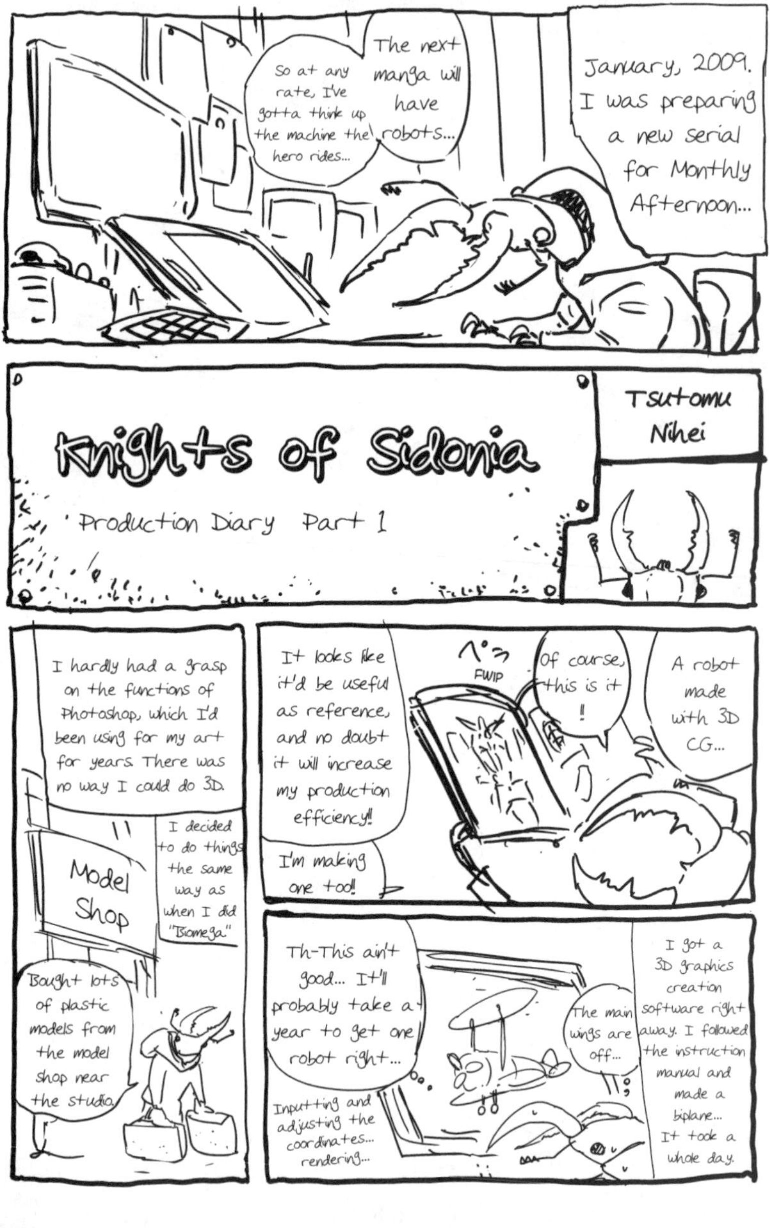
January, 2009. I was preparing a new serial for Monthly Afternoon...
The next manga will have robots...
So at any rate, I've gotta think up the machine the hero rides...
Tsutomu Nihei
Knights of Sidonia
Production Diary Part 1
A robot made with 3D CG...
Of course, this is it !!
FWIP
It looks like it'd be useful as reference, and no doubt it will increase my production efficiency!!
I'm making one too!!
I got a 3D graphics creation software right away. I followed the instruction manual and made a biplane... It took a whole day.
The main wings are off...
Th-This ain't good... It'll probably take a year to get one robot right...
Inputting and adjusting the coordinates... rendering...
I hardly had a grasp on the functions of Photoshop, which I'd been using for my art for years. There was no way I could do 3D.
I decided to do things the same way as when I did "Biomega"
Model Shop
Bought lots of plastic models from the model shop near the studio

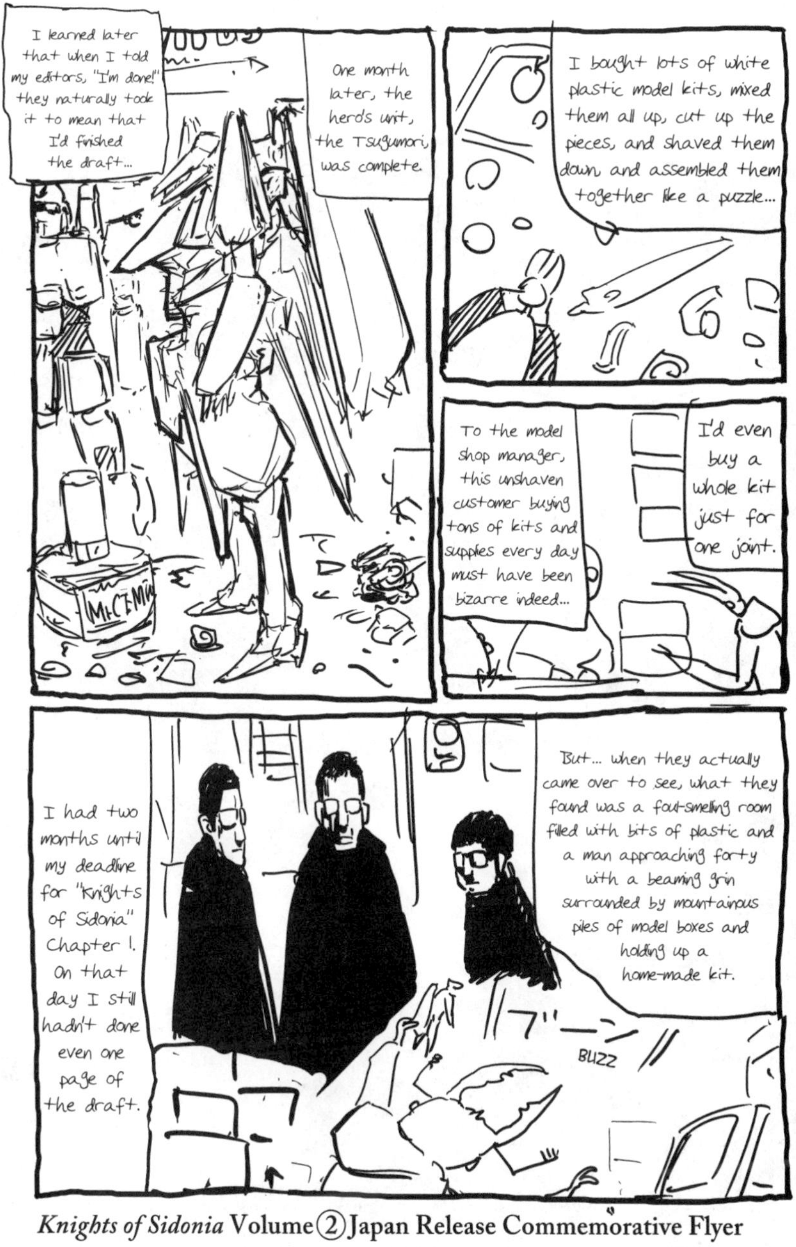
I bought lots of white plastic model kits, mixed them all up, cut up the pieces, and shaved them down, and assembled them together like a puzzle...
I'd even buy a whole kit just for one joint.
To the model shop manager, this unshaven customer buying tons of kits and supplies every day must have been bizarre indeed...
One month later, the hero's unit, the Tsugumori, was complete.
I learned later that when I told my editors, "I'm done!" they naturally took it to mean that I'd finished the draft...
But... when they actually came over to see, what they found was a foul-smelling room filled with bits of plastic and a man approaching forty with a beaming grin surrounded by mountainous piles of model boxes and holding up a home-made kit.
BUZZ
I had two months until my deadline for "Knights of Sidonia" Chapter 1. On that day I still hadn't done even one page of the draft.

Knights of Sidonia Volume ② Japan Release Commemorative Flyer

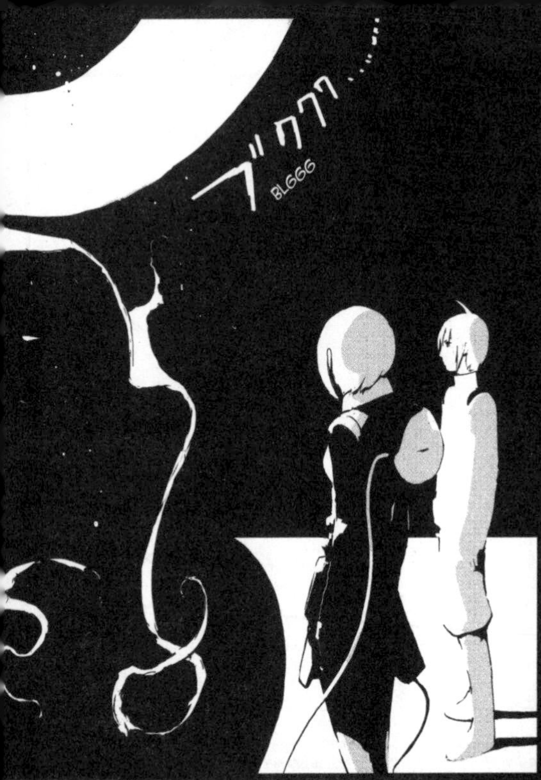
BLGGG

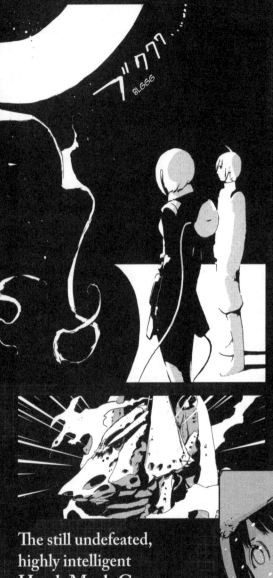

A new weapon that will change the face of the war,

"Gauna Core Piercing Rounds,"

excite the crew of Sidonia.

But they do not know.

That this weapon to destroy the Gauna originates from ghoulish research on Gauna.

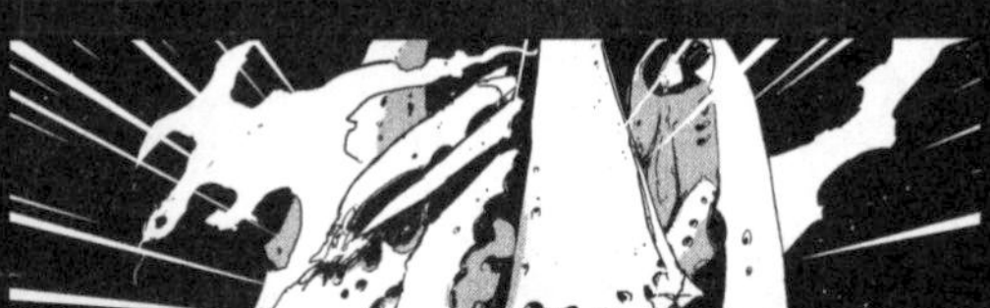

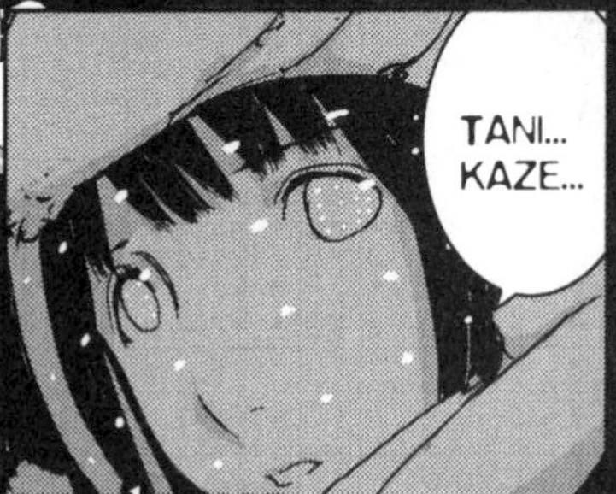
TANI... KAZE...

The still undefeated, highly intelligent Hawk Moth Gauna.

Why does "she," thought to possess human thought patterns, attack Sidonia?

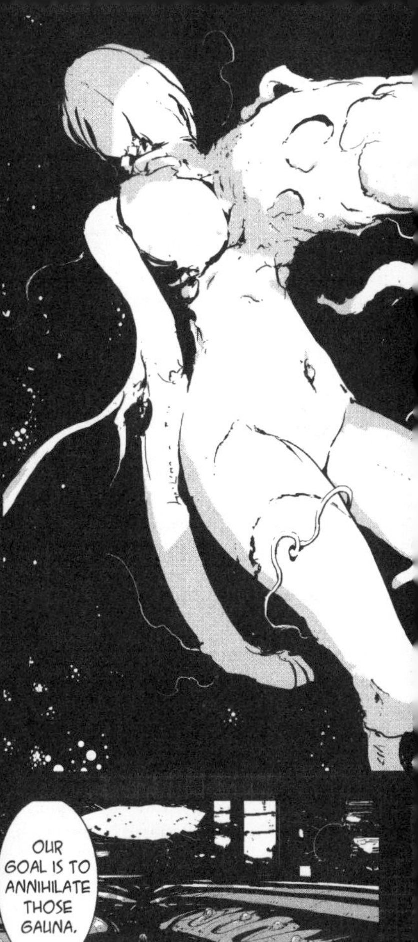

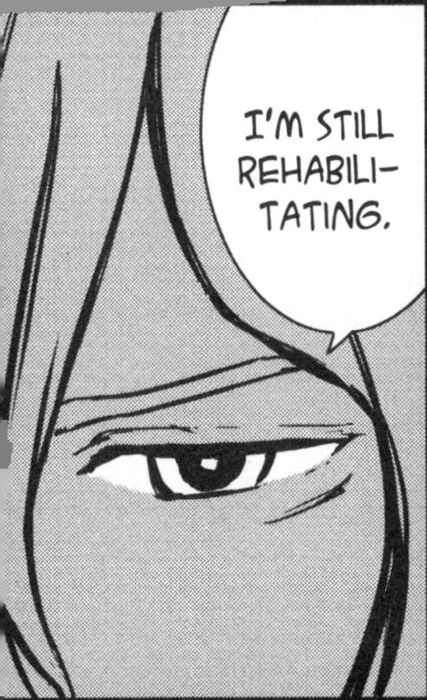
I'M STILL REHABILI- TATING.

What kind of power was behind Ochiai's seal, which Kunato has broken?

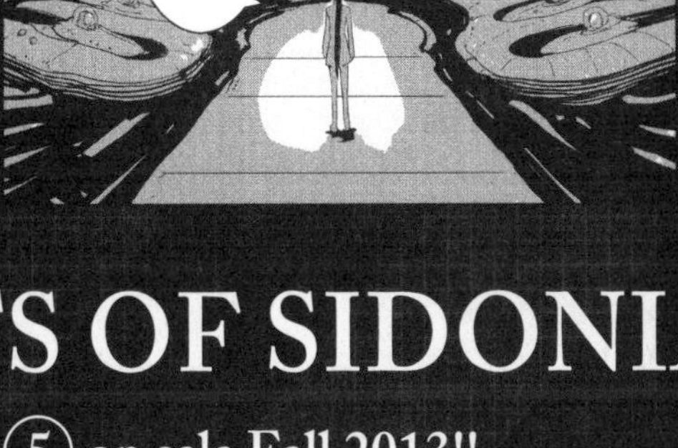
OUR GOAL IS TO ANNIHILATE THOSE GAUNA.

The battle strains to its absolute limits!!

KNIGHTS OF SIDONIA

Volume 5 on sale Fall 2013!!

Knights of Sidonia, volume 4

Translation: Kumar Sivasubramanian
Production: Grace Lu
　　　　　　 Daniela Yamada
　　　　　　 Tomoe Tsutsumi

Copyright © 2013 Tsutomu Nihei. All rights reserved.
First published in Japan in 2010 by Kodansha, Ltd., Tokyo
Publication for this English edition arranged through Kodansha, Ltd., Tokyo
English language version produced by Vertical, Inc.

Translation provided by Vertical, Inc., 2013
Published by Vertical, Inc., New York

Originally published in Japanese as *Shidonia no Kishi 4* by Kodansha, Ltd.
Shidonia no Kishi first serialized in *Afternoon*, Kodansha, Ltd., 2009-

This is a work of fiction.

ISBN: 978-1-935654-89-6

Manufactured in Canada

First Edition

Vertical, Inc.
451 Park Avenue South
7th Floor
New York, NY 10016
www.vertical-inc.com

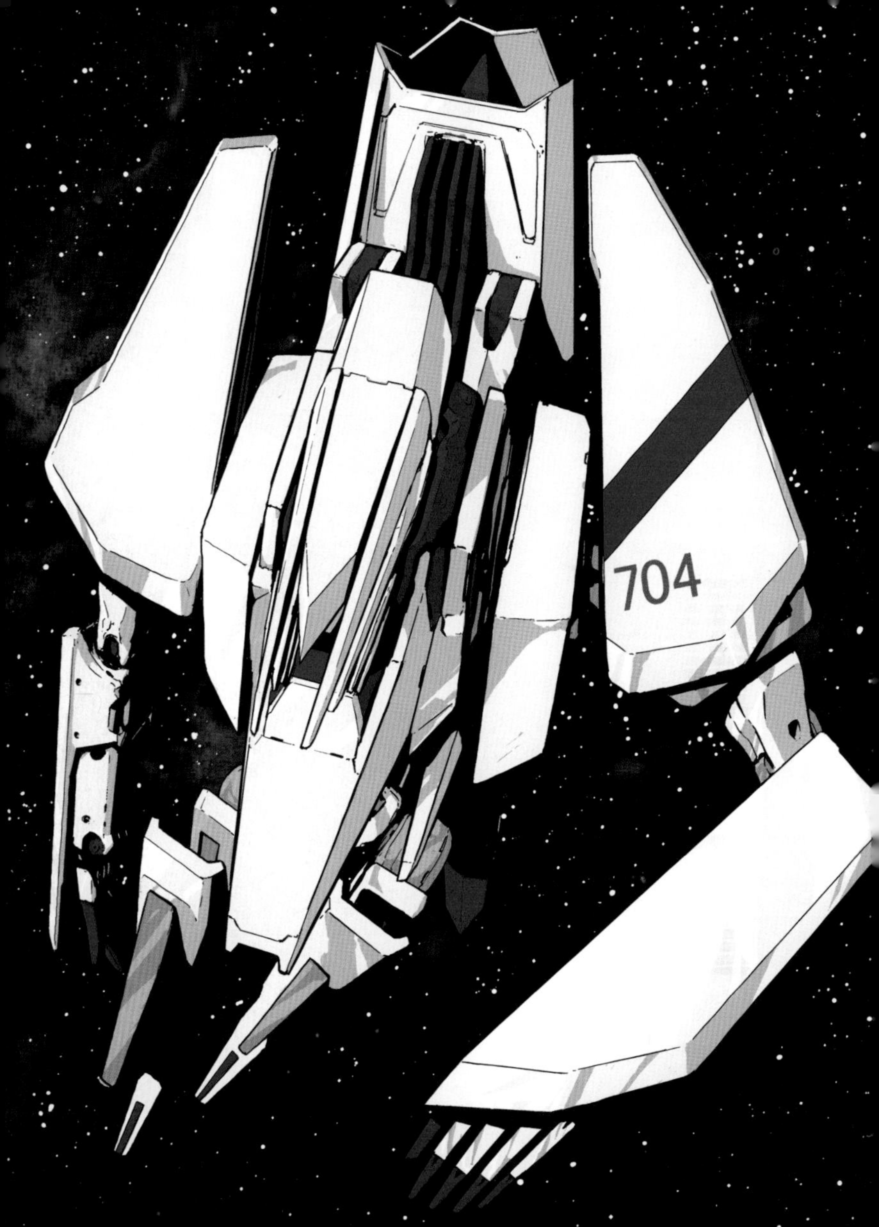
704

Supplement: Modified from the April 2010 cover of *Monthly Afternoon*

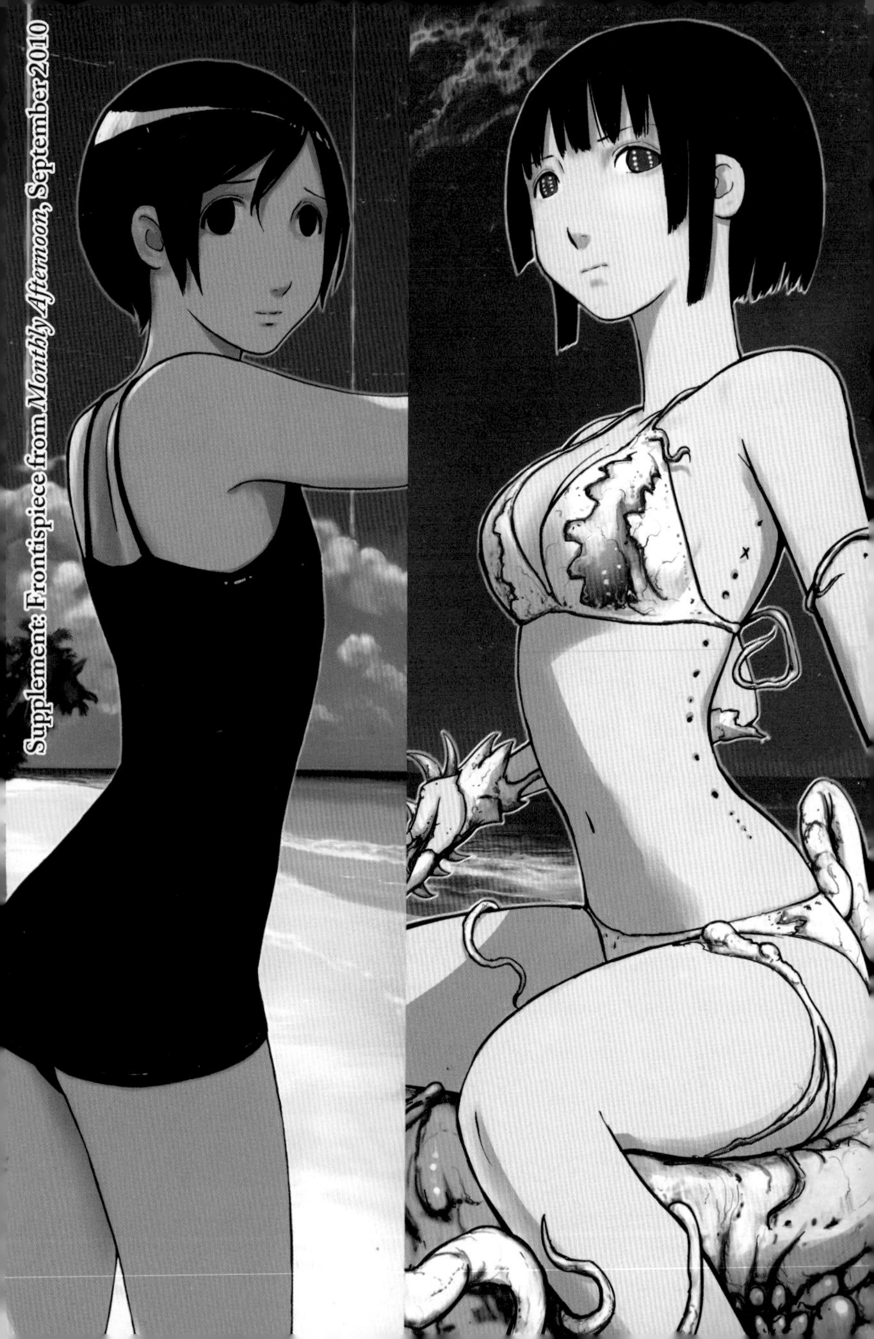

Supplement: Frontispiece from *Monthly Afternoon*, September 2010